Navigation Systems

for

Space Travel

by

Paige McCallister

and

Quantum Dancer

© 2020

Preface

The goal of this publication is to provide the best of navigational systems for the space traveler.

The authors have spent considerable time researching all things related to navigation for space travel. From this research and development period, we have developed best coordinate systems, best navigational tools, and best mapping programs, for traveling across space.

Thus, in the publication we provide the best of these navigational systems. We discuss the basic concepts of each system, the science behind each system, and the practical details of operation of each system.

Technical drawings are included.

The Need for Navigational Systems

We will be traveling across the galaxy. There is no doubt that we will do this. It is our future. We will explore and settle the inhabitable regions of space. We will commute regularly between various locations within our own galaxy....and beyond.

However, in order to make the space travel effective, we require several systems in place. This includes Navigational Systems. Therefore, in this publication we will describe the most effective navigational systems and related tools for long-distance space travel.

New Coordinate Systems

We begin with the Coordinate Systems. The existing coordinate systems are inadequate for the needs of the space traveler. Therefore we have developed two new coordinate systems.

The first is the Space Location Coordinate System. This system will designate the precise position of any object in space. This will be used by all navigation crew to locate specific planets, stars, and beacons during space travel.

The second coordinate system is the Universe Grid System. This coordinate system divides the entire universe into several Space Regions. This perspective is useful for considering all objects within a Space Region, as well as general oversight of each Region.

Navigational Tools

In this publication, we offer the following Navigational Tools for Space Travel: 1) Path Drawing Program; 2) Star-Map Room; 3) Space Beacons.

1. The Path-Drawing Program will trace the exact path the ship has taken from the home planet. This path can then be viewed on a Map.

2. The Star-Room Map is an immersive replica of the space environment. Actual photos of each region of space are displayed in the Dome around the viewer. The user of the Map-Room can therefore call up a specific region of space, and see this reality as shown around him.

3. The Space Beacon Network will be a collection of Beacons throughout the galaxy. Each beacon will act as a lighthouse, as it emits a unique signal. This signal will be matched to a Beacon Map, and therefore the navigation crew will know their exact location in space.

The Value of the Navigation Systems

Therefore, when using these Coordinate Systems, the Navigational Tools, and the Mapping Systems, the space travelers will be able to effectively navigate through the vast distances of space.

Using these Navigational Systems, all ships will remain on course, never deviating from their path. Such a direct route of travel will ensure the fastest travel between locations, and maximum safety for all travelers.

Each design is fully explained and illustrated in this publication.

Table of Contents

of

Navigation Systems for Space Travel

A. The Navigation Systems

B. Beacon Designs and Networks

11. Beacon Types and Designs

12. Drawings of Beacon Types and Components

13. Beacon Signal ID and Particle Supply

14. Other Supply Particle Capture Methods

15. Beacon Networks and Navigation

C. The Broad Perspective

16. Universe Grid System and Space Regions

17. Review of Navigation Systems

18. Glossary by Topic

Detailed Table of Contents

Chapter 1:

Introduction to Navigation Systems
for Space Travel

The Purpose of Navigational Systems

For any type of transpiration, a set of navigational systems is required. This includes a coordinate system, mapping systems, and tools to know your exact location. In total, any set of Navigational Systems must be able to answer the following questions:

1. Where are we now?

2. Where is our desired destination?

3. How do we get from here…to there?

In space, finding these answers becomes slightly more complex. However, using the right tools, the answers can be found easily. We will explain the tools and how to use them in the publication.

The Coordinate System for Space

The first objective in designing the Navigational Systems is to create an effective Coordinate System. We need to create a consistent, and accurate, method for designating the location of every object in space.

We will be using a form of Cartesian coordinates, yet for three dimensions, and for all of the universe. This will be similar to the longitude and latitude system used on Earth, yet for locations in deep space.

The center point will be Earth. The angles and distances will be measured from Earth. Note also that the specific mechanics of the measurements are important. Therefore we will describe the measuring system in detail.

Once we have established this Coordinate System, we will use it to designate precise locations for every object in space. Traveling ships will use the same Coordinate System to know their current location.

Navigational Tools:
Knowing Your Location

The space traveler will want to answer those questions posed above. These questions include: Where are we? Where is our destination? How do we get from here to there?

The navigation crew will be able to answer these questions using a set of navigational tools. Some of these are maps and databases. Other tools are measurement equipment on board the ship. The most important of these tools will be the navigational beacons throughout the galaxy.

The majority of this publication will describe the Navigational Systems for determining your location. In brief, these include the following:

- Visual Sighting
- Path Drawing Program
- Space Beacons and Beacon Map
- Star-Map Room

Visual Sighting

The Visual Sighting method is the same method used by travelers for centuries. The traveler uses a set of stars, relative to the direction he wants to travel. As long as he keeps certain stars to his left and right, then he should be traveling in approximately the correct direction.

Of course, this method has many limitations. Therefore we have developed several other navigational tools, each of which will provide greater accuracy.

Path Drawing Program

The Path Drawing Program is a tool which draws the path taken by the ship. This Program is actually a collection of tools. The specific tools include the mechanics of measuring the ship's movements; several computer calculation programs; the databases; and maps.

In brief: whenever the ship changes direction (X,Y, or Z) the Path Drawing Program makes note of this. Whenever the ship changes speed, the Path Drawing Program also makes note of this. Therefore, whenever the navigational crew looks at the Final Map, they will know their location, relative to their Home Planet (of Earth).

Mapping Systems

In any form of transportation, maps are essential. Therefore, the Space Corps will offer several Mapping Programs. Each Mapping Program will be sophisticated, and be able to serve the needs of those who travel across the galaxy. Note that each Mapping Program will serve a different need.

The most important of these Mapping Programs include:

- Path Drawing Maps

- Beacon Location Maps

- Star-Map Rooms

We will begin with short overviews of each. Additional details are provided within the relevant chapters. Furthermore, a separate publication is devoted just to this topic.

Path Drawing Maps: Overview

The Path Drawing Maps will show the exact path taken by the ship. This will be shown in two main ways: Path Steps, and Direct Path.

The Path Steps are the specific paths taken by the ship during the journey. Each time the ship changes direction or changes speed, this is registered as a new "Event. On the map, this will be placed as dot. Thus, each dot represents a specific direction and speed. The computer will then draw lines between these dots. These lines represent the Path Step between the two Events. Therefore, we can view the entire journey of the ship as a series of Path Steps; each of which is shown on the screen as a series of short lines.

The Direct Path is the direct path taken from the home planet to the current location. Specifically, this is the Direct Angle (XYZ angle) and the Direct Distance, from home planet to current location. The computer will calculate these values from the series of Path Steps above. Thus, the computer takes each of the Path Steps (with all the angles, speeds, and distances, to create the shortest route). The computer will produce the Direct Angles (X-Y and Z angles) and the Direct Distance, from home planet to current location. These values can be used for official locations, such as the location of the new Space Beacon. The result can also be mapped, shown as a direct line on the screen.

Beacon Location Maps: Overview

The Beacon Location Maps will show the locations of every Space Beacon in the network. These Beacon Maps are essential mapping programs, as they will be the main navigational tools for the ships.

The Space Beacons emit bursts of EM energy, similar to a lighthouse. The emission from each Beacon is unique, and thus is a Beacon ID Signal. When a ship passes by the Space Beacon, the ship will receive the specific signal, and be able to match the ID Signal to the Beacon Number. The crew can then match the Beacon they are near to the point on their Beacon Map. This will confirm the ships location (or help the crew adjust course as needed).

The Beacon Map itself is a computer program with the official location data for all the Space Beacons. The data includes Direct Angle from home planet; Distance from home planet; Beacon Signal ID factors, Name of Beacon Line, and Specific Beacon Number. The crew than then view this data in various ways, which includes various types of maps.

Star-Map Room: Overview

The Star-Map Room is an immersive mapping system. The room has a dome display, where stars are shown on the screen around you. Thus, when you stand in the room, it is as if you are looking at the stars which surround the ship.

The Dome Display works as a series of screens, similar to digital display screens. There are numerous most sized screens (rather than one big screen), each approximately 2 feet per side. Each will display its own image. Then, taken together, the Dome Display will show all of the stars, in all the positions.

Star-Map Room: Viewing Options

This immersive mapping system can be used for a variety of purposes. There are main viewing options are Live Feed and Remote Location View.

Live Feed in Star-Map Room

The images will begin as Live Feed, from cameras on the top and bottom of the ship. Therefore, the crew can see the entire area of space around them, when inside the Map Room.

Photos Taken During the Travels

Furthermore, these images will also be captured as screen shots, and stored in the computer. These stored images will be uploaded to future ships, and used in the Star-Map Room in various ways. Therefore, the most common viewing option for each successive ship to travel the route will be the Remote Location View.

Remote Location View

The "Remote Location View" will allow you to choose other locations in space, far away from your current location, and see what the stars will look like from that location. These images are again taken from still images recorded during previous journeys. Yet now you can see ahead.

Therefore, you can choose a few stars, tell the computer which part of the dome to display, and the mapping system will fill in the rest. You can get an immersive preview of what the stars will look like, when you get to that position in space.

See Publication on Maps for Details

These descriptions are just the brief overviews for each Mapping System. To learn the full details of each Mapping System, please see the separate publication: "Databases and Mapping Systems in the Space Corps". In addition, some of those details will be presented in relevant chapters in this publication.

Chapter 2:

The Coordinate System
for
Locations in Space

Introduction to Space Coordinate System
and Importance of this System

If we want to create any type of mapping or navigation system, we must first create a proper coordinate system, with a common reference point. This system must also work effectively for all locations in space.

The following system has been carefully created to meet each of those requirements. Using this system we can accurately know the locations of all Space Beacons, all planets, and all other celestial objects. This system will be used for all mapping systems and navigational systems, and therefore also be used by anyone who travels across the galaxies.

Official Locations for Objects in Space

The Official Locations for any object in space will be given by the following values:

- Direct Angle: X-Y Angle and Z Angle
- Direct Distance: as Measured from Earth

These parameters will be able to precisely state the location of any object in space. This includes: remote planets, stars, space beacons, and other celestial objects.

Overview of the Coordinate System
for Locations in Space

The Coordinate System we will use is based on the Earth as the reference point, with the angles and distances defined according certain aspects of the Earth. Specifically, we will know the precise location of any object in space using the following parameters:

1. Reference Point = Earth as Home Planet

2. X Angle = Earth's Orbit Position;

   X = 0° where Earth is closest to sun
   Numbering proceeds in direction of Earth's Orbit

3. Z Angle = Angle from Equator of Earth

   Z = 0° at Equator

   + Z is angle toward North Pole
   - Z is angle toward South Pole

4. Distance = Distance from Earth
   Earth's orbit point where Earth is closest to object

Therefore, using these specific parameters, we can specify the exact location of any object in space. We can give precise locations for each Space Beacon, each Planet, and each Star in the galaxy.

Each parameter in this coordinate system is further discussed and illustrated below.

Reference Point: Earth

All location systems and mapping systems require a reference point. This will make our databases and maps consistent, regardless of where the objects are located in space.

The official reference point will be Earth. This is because our civilization will be starting from Earth. Therefore all zero values will be based on the Earth. For the residents on other planets, the computer programs will be able to convert these official values to the perspective of different home planets.

Earth Closest to Object

The Earth orbits the sun, therefore it is not in one place. The distance between the Earth and a distant location will vary accordingly. Therefore, to establish consistent distance values, the zero distance for the Earth will be designated at one location.

There are two options. The first option (the choice we will use) is where the Earth, in its orbit, is "closest" to the object. This is most reasonable, as this is the Earth's position when ships will embark. Therefore, for practical reasons (of interest to the actual travelers), this is the zero distance.

The second option is to use the position of the Earth where the Earth is closest to the Sun, as the zero for all distance locations. We will also include this value in the database.

However, the first option will be the beginning point for all ships, and therefore will begin the official distance measurements.

Distance to Location in Space

The distances to each object in space will be measured from the Earth to the object. We will measure the distance to all objects from the Earth. This is the only sensible option, as Earth is our first home planet. The distances will be measured along the X,Y,Z angles from Earth to the destination.

Specifically, the distance will be measured from the orbital position of the Earth which is closest to the object. The precise distance value is measured based on the position of the Earth's Orbit. Specifically, as the Earth orbits the Sun, there is one position where the earth is "closest" to the settlement planet or Space Beacon. It is this location of the Earth where we set the zero distance.

The reasoning for this point as the zero distance is simple. This is the location where all ships will embark when traveling that direction. It is also the direction between the two planets. Therefore, this is the most practical choice for zero distance.

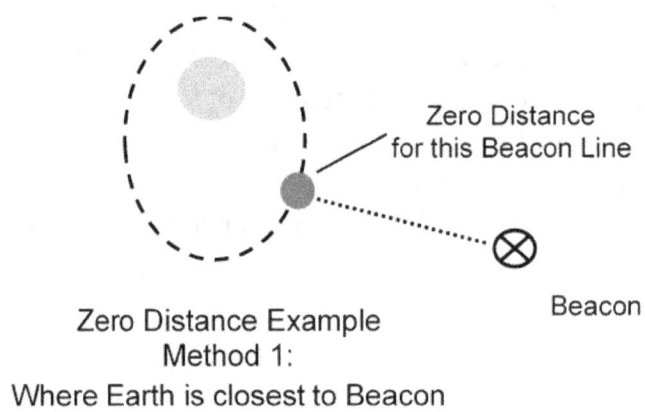

Zero Distance
for this Beacon Line

Beacon

Zero Distance Example
Method 1:
Where Earth is closest to Beacon

Therefore, the "distance" value will always be taken from Earth, where the Earth is closest to the Space Beacon or Planet. This is how the distances for all objects will be measured, including the distances to Space Beacons, Stars, and Settlement Planets.

Of course, the Earth will be at a different location on its orbit for the zero point, depending on the direction of the planet or beacon. However, for practical purposes of space travel, this is the most sensible approach.

Alternate Zero Distance:
When Earth is Closest to Sun

Note that an alternate location is to place the Earth at the closest point to the Sun (this is also the 0° for the X-Y angle), for all distance measurements. This would provide absolute consistency for distance measurements in all directions.

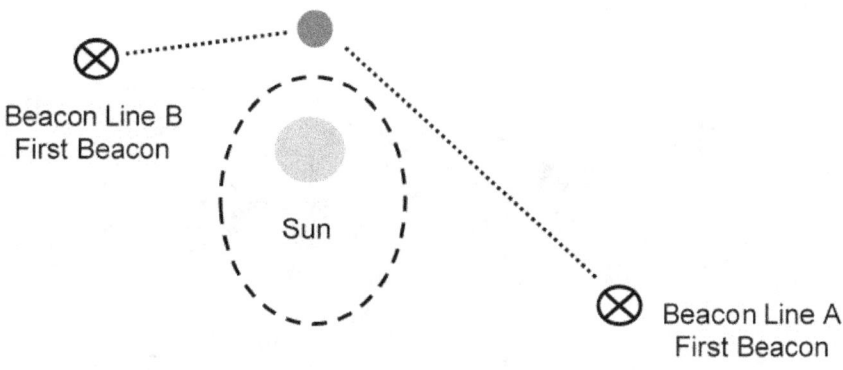

Beacon Line B
First Beacon

Sun

Beacon Line A
First Beacon

Zero Distance Example
Method 2:
Where Earth is closest to Sun

However, while this is more universal, it is less practical. For practical measuring, and using the Path Drawing Program, it is better to use the system where the Earth is closest to the remote object. Therefore the official distance measurements should be based on the where the Earth is closest to the remote object. This is of most use to space travelers.

Then a secondary list where the distances are modified for the zero position of the Earth on the orbit plane. This will be of use to the scientists, but not as much use for the space travelers. This will be an additional distance measurement, but not the official distance measurement.

Both Distances in the Database

If desired, we can list both distances in the database. We will use the position of the Earth when closest to the Beacon or Planet as the zero distance for our Path Drawing Program, and thus for our official values. (This is the process described throughout the publication).

Then, the computer will perform appropriate calculations to list the same distances when the Earth is a the 0° for the X-Y angle. These dual distances can be listed in the database, and the user can then choose which values he prefers.

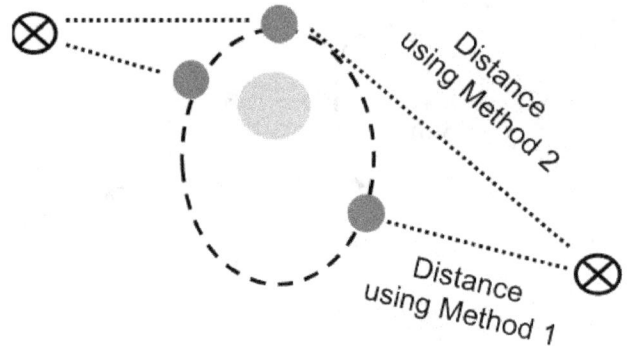

Directions: X, Y, Z Angles

Every object in space, whether Space Beacon, Planet, or Star, will be a specific Angle from Earth. This Angle must be noted in two directions, the X-Y angle, and the Z angle.

The direction to any space beacons and settlement planet will listed as angles in the X, Y, and Z directions, as based from Earth. The X-Y angles can be visualized as a compass circle. This circle is laid on the same plane as the Earth's Orbit.

X-Y Angles, from Earth to Any Space Location

The X-Y angle is essentially the orientation on a flat plane. The X-Y angles can be visualized as a compass circle. We can use the plane of our Solar System for this purpose. Thus, as the Earth orbits the sun, the "angle" from Earth to another object will be of the X-Y angle.

The zero degree position is something we will impose. This could be located anywhere on the Earth's orbit. However, the ideal location for the zero degree X angle is where the Earth is closest to the sun.

Thus, the Zero degree of the X-Y circle is designated as the position of the Earth where the Earth is closest to the Sun. The degree values from 0 to 360 then proceed in the same direction as the path of the Earth's orbit.

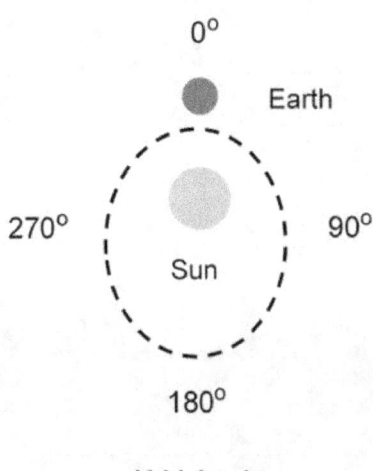

X-Y Angle
Coordinate System

Z Angles, from Earth to Any Space Location

The Z angle is the vertical angle from the Earth's equator. Because the X-Y circle sits along the plane of the Earth's orbit, the Z angle is the up and down direction from that plane. Specifically, the zero angle for Z is the equator line of the Earth. A "positive" Z angle raises upward from the Equator, while a "negative" Z angle lowers below the Equator.

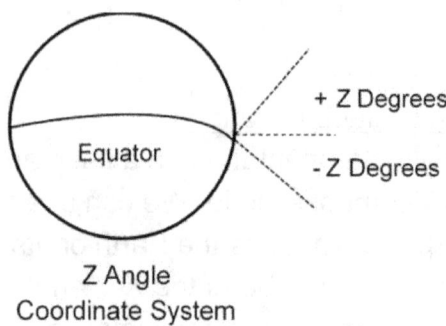

Z Angle
Coordinate System

Adjusting for Solar Movements

For most practical purposes, we can consider stars as stationary. However, stars do move, and we will eventually need to account for this. Thus, every 50 years we will make adjustments in our database values.

As a simple example, if our sun moves away from a Beacon Line, then the distance to each Beacon will increase. Similarly, the sun may be moving toward another Beacon Line, in which case the distance to each Beacon on this line will decrease.

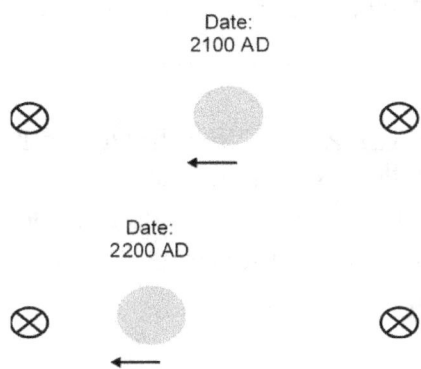

Furthermore, other settlement planets have their own stars which move, and change their distances to each Beacon. And, of course, with both of our solar systems moving...the combination of movements will produces changes in distances and angles between each settlement.

These changes will be small from year to year, but will accumulate over the decades and centuries. Thus, if not checked, if not kept up with, the shifts in reality will become far from accurate in the database and maps. A few updates, every few decades, will be enough to keep everything updated in the database.

The details for the adjustment process are explained in other publications. In brief, the process begins with monitoring the movements of the solar systems. When location data requires adjustment, the scientists will then use their computers to calculate the exact proper values, for each Star, Planet, and Beacon in the databases.

These adjustments can be as simple or as complex as desired. The ideal of course is to maintain the highest accuracy. (See other publications for specific processes).

Settlement Planets:
Dual Reference Points and Location Conversions

In the far future, we will develop Dual Reference Point Systems. As we extend further into space, the settlers will want navigational systems based on their reference. When this time comes, we develop a second reference point system.

The settlers will adjust all the values for the Database from Earth's perspective, to be from the perspective of their own home planet. (This is discussed in greater detail in the Publication on Databases and Mapping Systems).

Conversions of Angles and Distances from Different Planet

In addition, each Settlement Planet will have the computer program which can convert the angles and distances to their needs. They select a new Home Planet from the computer program, then the computer automatically recalculates the values.

Official Location Points Always from Earth

However, the Earth as reference point will *always* be the primary reference point, in all databases and maps. This will be the official value, regardless of where the user is living. This is essential for consistency in navigation and mapping systems for all ships, wherever they travel.

Thus, the ship can be anywhere in the galaxy, yet the locations will always be officially designated by the measurements from Earth. The ship, and the remote planet, can of course convert a specific set of data for their purposes, using the programs mentioned above. Yet the Official Locations, for all objects in space, will be designated using the measurements from Earth.

Review of Coordinate System
for Location of All Objects in Galaxy

Before we set up any Space Beacons, and before we embark on any space travel, we must first establish a proper system for knowing the exact locations of any object in space. This system must be simple, and yet be able to work for the location of any object in the galaxy. The best system is to have the four parameters:

1. Reference Point

2. X-Y Angle from Reference Point to Object

3. Z Angle from Reference Point to Object

4. Distance from Reference Point to Object

Given these four essential parameters for specifying locations, we have selected the ideal choices for those parameters.

1. The best reference point is that of planet Earth. This is because our civilization begins on Earth. We are embarking from Earth. Therefore, it is sensible to make Earth as the Reference Point for all other values.

2. The X-Y angle coordinates is based on a plane in space. The best plane to use is the Solar Plane; specifically the orbital plane of planet Earth. Furthermore, as the Earth orbits the sun, so the angle degrees circles from 0 to 360. The zero point can be placed anywhere. However, the ideal location is where the Earth is closest to the sun. The degrees will then increase in the direction which the Earth orbits.

3. The Z angle is the perpendicular angle to the chosen X-Y plane. This means that the Z angle is vertical inclination from the Earth's orbiting plane. Thus, the +Z is upward from the solar plane, and the -Z is downward from the solar plane.

However, for practical measurements (see Pathway Drawing Program) it is best to choose the equator of the Earth. Thus, the +Z is best measured as the inclination from the equator toward the north pole, while the -Z is measured as the declination, from equator toward south pole.

4. The Distance is measured from the Reference Point to the Object. This means we start our zero distance at the location of the Earth.

However, the Earth does orbit, and the space object can be in any direction from the Earth. Therefore, we must specify the location of the Earth, in its orbit, when we start the zero distance.

The best method is to use the location of the Earth, where it is closest to the object. This is the simplest and most accurate, as this is the position of the Earth when all ships traveling that direction will embark. This is also the position of embarkation for the Beacon Placement Ship. Thus, this location is the preferred location for the Earth as zero distance.

Of course, this also means that the Beacon Lines in each direction will use the Earth in a different orbit location. Yet, again, this is the point of embarkation for all ships traveling in that direction. Therefore, from a practical use point of view, these are the distances of interest to the ships. The differences in starting location is not as important, for the ship is not traveling in those other directions.

4b. If desired, we can choose a common location of Earth's orbit, for the zero distance. The best common location for this purpose is the same as the zero angle for the X-Y; which is where the Earth is closest to the sun. If this is desired, it is best to use both locations in the data. The distance is most accurately measured when starting from the embarking position. These values can then be converted to the other starting location.

Therefore, in total: the specific parameters, using Earth's positions as described above, will always be the coordinate system for locating all objects in space. These parameters will apply to all Space Beacons; all Planets; and all Stars. This system will be applied consistently across the entire galaxy and beyond.

Indeed, this Coordinate System will allow us to create effective navigation and mapping systems, for all activities of the Space Corps.

Chapter 3:

The Path Drawing Program

Short Overview

The Path Drawing Program is a mapping system which tracks the precise movements of the ship. The exact path of the ship is then drawn on the computer screen. Therefore, using this Program, the navigational crew will be able to know their exact location, relative to the Home Planet.

This navigational tool will be used for a variety of purposes. The most important purpose is for Beacon Placements. The Program can also be used to know their location in deep space.

Path Drawing Program and Space Beacons

The Space Beacons are the main navigational tool of the space traveler. Therefore, it is important to know the precise location of each Beacon. Thus, the Path Drawing Program will be used primarily by the Beacon Placement Ships, when placing the Beacons.

The Beacon Placement Ships will be the first ships to travel this route between planets, and therefore will also place the Beacons into position.

The Path Drawing Program on these ships will therefore be the most sophisticated and accurate, of any such programs, on any ships. The crews of these ships will therefore know the precise location of each Space Beacon. This data becomes entered into the Beacon Database, and becomes the Beacon Map.

Therefore, in total: the Path Drawing Program, as used by the Beacon Placement Ships, will determine the official location of each Beacon. All future ships will have the latest version of the Beacon Map, as a Navigational Tool during space travel.

Note that the details of the Space Beacon operations, as Navigational Guides, will be discussed in later chapters.

Path Drawing Program as Used by Any Ship

The basic design of the Path Drawing Program can be used by any ship. However, the accuracy depends on the sophistication of the engineering, and the precision in assembly. Therefore, we recommend that only the Beacon Placement Ship have the most accurate Path Drawing Program. All other ships will have simplified versions. Note also that most ships will use the Beacon Map and Star-Map Room for navigation, in conjunction with their Path Drawing Program.

General Overview

The Path Drawing Program can be compared to a long colored string, or a line of paint. Imagine a long roll of colored string. As the ship moves forward, the string unrolls. Thus a path is traced, in space, from Earth to the current destination. Every turn, in any direction, the ship leaves this string as evidence. This is what the Path Drawing Program does.

We can also visualize this as a line of paint, gently brushed down behind the ship as it travels in space. Thus, again, where the ship goes, in any direction, the line of paint is left behind as the trace of the path. Again, this is what the Path Drawing Program does.

Although there is no actual string or paint being released behind the ship, we do create that same type of Colored Line, on a sophisticated computer program. This is how we obtain the precise location of each Space Beacon that we place into position.

The Path Drawing Program is able to take all information of the ships travel, in any direction, over the entire journey, and trace a path on the computer screen. Every turn, every change in angle, every kilometer traveled, the Program knows.

This data is measured, recorded, and analyzed. From this we create several lists, and several maps. The ultimate map, and the most important, is the map of the Direct Line between two points, which tells us the exact position of the Beacon relative to the Earth.

Therefore the Path Drawing Program is the essential tool for determining the precise location of each Beacon we place along the route.

However, in order to make this work, we need very sophisticated mechanical engineering technology. We also require sensitive data acquisition devices, and complex computer programs. This means the most expensive and sophisticated components available, in each of the related engineering areas.

Note also that we group these technologies together, under the singular term "Path Drawing Program". This "Program" is not only the computer program which creates the final maps, but also includes the mechanical engineering to detect and measure the ship's motion, and other related devices.

The details of the Path Drawing Program are now presented below.

Multiple Components and Programs

The Path Drawing Program is an essential tool for knowing the precise location of the ship, relative to the Earth.

Note that the "Program" is not just one program, but rather is several interrelated programs. Furthermore, and this is very important, the "Program" involves extremely sophisticated mechanical engineering.

Therefore, all of these sophisticated engineering tools must be built into the ship. These tools are essential for knowing the exact location of the ship, and therefore the exact location of the Space Beacon.

We must emphasize again: it is from this Path Drawing Program, with all related mechanical engineering and computer programs, that we know the exact location of the Beacon. This is the Tool which determines the official location of each Space Beacon.

In the following sections we will discuss the details of this important navigational tool.

Path Drawing Program:
User Perspective

The Path Drawing Program is the primary tool for knowing the exact location of the Space Beacons. The "program" is a set of mechanical and computer tools. Taken together, this Program will eventually produce the Direct Angle and Direct Path between the home planet and the specific Space Beacon.

Most of the operations are automatic, however some of the operations will involve the crew. These operations will be summarized here, from the perspective of the crew user.

Events in the Program

As the ship travels through space, the mechanical devices note the angle and speed of the ship. Each time the ship makes an adjustment, a new "Event" is created in the database. An "Event" includes any of the following: change in X angle, change in Z angle, or change in speed. The user can look at this list of Events at any time.

Path Steps

These Events are also endpoints of the "Path Steps". When any two Events have occurred, the computer can calculate the total time, and thus the total distance. The data between the end points will also create the "Path Steps". These Path Steps can be mapped as segments, where each segment is a Path Step. The user can view the entire series of Path Steps at any time, to see the history of the ship's travel.

Direct Path

The final step is the "Direct Path". When the crew stops at the new location to place the next Space Beacon, they will have the computer perform the final calculations. The computer will combine all of the individual Path Steps, to calculate the Direct Path. This is the shortest line between the home planet and the current location. The Direct Path can be drawn as a single line in the mapping program.

Direct Path for Official Location of Space Beacon

The values calculated for the Direct Path will be the Official Values for the Space Beacon Location. Thus, when the crew is ready to place the next Space Beacon, they instruct the computer to perform the final calculations. This will give the values of the Direct Path, which include: Direct X Angle; Direct Z Angle; and Direct Distance. These values are then manually entered into the Beacon Location Database.

Path Drawing Program Database

The Path Drawing Program is the tool which accurately measures the path taken by the ship. However, the "Program" is actually not just one program, but a set of connected tools. Several of these tools are specific databases. We will therefore discuss those databases in this section.

Regarding specific data, the primary purpose of the Path Drawing Program is obtain the precise Direct Angle (X and Z), and Direct Distance between the home planet and the current location.

Some of this data is generated automatically. However, the final values, for the Direct Distance and Direct Angle, must be selected by the crew operator. That is, the crew tells the computer to calculate.

The crew can also review this data at any time, ideally at the beginning of their 8 hour shift.

In this section we will review some of the main concepts, especially those related to the databases of the program.

Data Collection in Path Drawing Program

As a reminder, notice that there are essentially three steps in the process. Each process is updated in different ways, using different computer programs, at different times.

1) The first process is the collection of raw data regarding the travel. This data includes the XYZ degrees of travel, and speed of the ship. This update is automatic, and will occur whenever there is a change in direction or change in speed. This data is sent to the computer after the change has become stable for 20 minutes. (Stability is important so the raw data is only the final values, not values during the change). This is recorded as a series of "events", each with a date-time stamp. This list of events becomes the first database.

45

2) The second program analyzes the raw data, to determine the total time and total distance traveled in each direction. The computer program will update this information automatically every 8 hours. This list of data becomes the second database. In addition, the user can choose to view the information as a map of travel history.

3) The final step is the calculations and maps, for the Direct Angle and Direct Line between home planet and current location. These calculations and maps are only activated by the human operator, and only when they are actually ready to know their location. This step will mostly be used when the Beacon Crews place the next Beacon into position.

Note also that this Official Data will be manually entered into the Beacon Database, then sent to all appropriate Corps Staff.

Automatic Data Acquisition

Notice that the data for each of these databases will be acquired automatically. There will be no human data entry in any of these processes. The mechanics and electronics will do this for you. The human operator then has choices on how to view the data of each database.

See Other Publication for Details

The reader will find additional details on the measurement and calculation processes for the Path Drawing Program in other publications. The main publication with this information is "Databases and Mapping Systems in the Space Corps".

The user operations will be discussed below.

Path Drawing Maps:
Path Steps and Direct Path

The "Path Drawing Program" exists to collect data and create maps regarding the ship's travel. The primary goal is to create path segments on the computer screen which represent the travels of the ship. This will allow us to calculate the Direct Path, between the Home Planet and current location. The final result, will then be the exact Distance and Angles between our Home Planet and current position.

This information can be shown on several maps. The two main mapping systems associated with the Path Drawing Program are the Path Step Map, and the Direct Path Map. These can be displayed individually, or both simultaneously.

Path Step Map: Events and Paths

The "Path Step Map" shows the exact paths taken by the ship. On the map, each segment represents the distance taken along a specific trajectory. In this viewing, each dot represents a change in direction or speed. Each line represents the distance traveled from that change.

Change of Course Event

Each time the ship changes course, it is recorded as a "Change of Course Event". This is often referred to more simply as an "Event". These "Change of Course Events" are listed in the Path Drawing Database.

These "Change of Course Events" primarily exist as a list within the computer. However, the Events can also be shown on the map, where each dot on the screen represents the location where the ship adjusted course or speed.

Path Steps

Each "Path Step" is the distance the ship traveled, along the precise XYZ angle, before changing course again. These Path Steps can be viewed as line segments on the map. The Path Steps are automatically updated every 8 hours.

Direct Path Map

The "Direct Path" is the direct line, from the home planet to the current location. Most often, this is used to show the direct path between Earth and each Space Beacon.

The Direct Path is measured as specific angle, in X and Z, with a specific distance. This is the Direct Angle and the Direct Distance, from Earth to the current location.

This is most often used for the determining the precise location of the next space beacon, when the beacon is placed into position. However, a crew can use this program at any time to know their current location.

Calculating Direct Path from Path Steps

These values are determined by the Path Drawing Program. The tools of the program first compute the series of Path Steps, as described above. The computer will then use all of that information, to precisely calculate the exact angles (X & Z), and exact distance, between Earth and the Space Beacon. These values will be clearly shown on the computer screen, so that the crew will know them for the Beacon Database.

Direct Path as Viewed on the Mapping Program

In addition, this Direct Path can also be mapped in the Path Drawing Program. This Direct Path can be shown along with the series of Path Steps, or shown independently.

Example Map of Path Drawing Program

The map below shows the paths of the ship, from Earth to one of the Space Beacons. The black line segments are the individual Path Steps. The dashed blue line represents the Direct Path

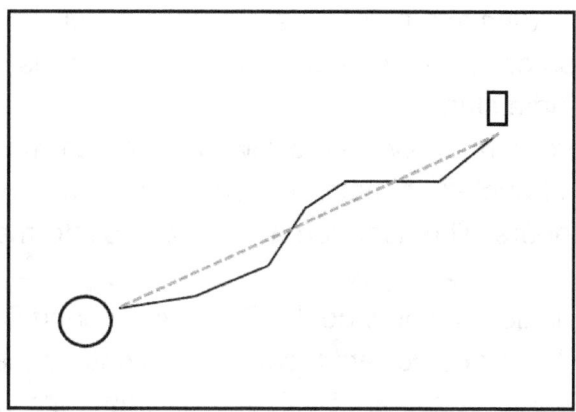

In the drawing above, each black line segment is a specific path step. There are dots (not shown, yet exist in the real map program) which are the locations where the ship changed trajectory or speed. The short segments between each dot is the distance traveled along that specific course. Each of these segments are Path Steps.

The "Direct Path" is the straight line between the end points of the Path Steps. (This is usually between home planet and a beacon). This values of are the Direct Angle (X, Z angles) and Direct Distance to the Space Beacon, which are then added to the Beacon Database.

Computer Image Manipulation

Note that this example is shown only in two dimensions. However, the actual path of the ship will be in three directions. The computer will be able to show the Path Steps in all directions, using different colored lines for X vs Z directions. The Path Step Map can also be rotated in different directions. There are many computer programs capable of these options.

A menu system on the side can allow user to change views, select different points, and show exact data values. The user can also zoom in or out as needed. Distance will be given in kilometers.

Reminders of the Updating Processes

As a reminder, notice that there are essentially three steps in the process for the final data and path drawing. Each step is updated in different ways, using different computer programs.

The first process is the collection of raw data regarding XYZ degrees and speed. This update is automatic, and will occur whenever there is a change in direction or speed. This is recorded as a series of "events", each with a date-time stamp.

The second program analyzes the raw data, to determine the total time and total distance traveled, in that direction. This will be automatically updated every 8 hours. The user can also view this information as a map of travel history.

The final calculation will provide the Direct Angle and Direct Distance, between Home Planet and current location. This data is the ultimate goal of the Pathway Drawing Program System. This final data can also be shown on a mapping program, as a useful visual aid. Note that these calculations and maps are only activated by the human operator.

Additional Technical Details
of the Path Drawing Program

The following sections will provide additional technical details of the Path Drawing Program. These details are related to the mechanical engineering and computer programs.

Note that the concepts presented here are somewhat brief. These details are explored in greater detail in other publications. Please see those other publications for the technical details regarding the actual design of components for this system

Mechanical Engineering for Knowing Direction and Speed

The first set of technologies required are the methods for determining direction and speed. These technologies must be extremely accurate.

Therefore, the first technologies for the Path Drawing Program System are those which can measure all direction angles and speed of the ship as accurately as possible. This will provide feedback to the crew, to ensure that they are piloting the ship as desired. This information is also important for the databases and mapping systems of the Path Drawing Program.

The accuracy of these components will depend on the sophistication of the engineering, as well as the precision in manufacturing and assembly.

Computer Database Will Record Raw Data, for Each Step

The first computer program will collect the raw data regarding ship's travel. This information will be recorded in a long series of steps. This information will also be collected automatically.

Specifically, each time the ship changes direction or changes speed, this computer will begin a new line of data. Thus, any change in direction or change in speed will trigger this program.

However, the program will also know not to start its new data entry until the change is again at a steady state. That is, the entry will begin when the ship has pivoted to its new direction and remained there for at least 20 minutes. Regarding speed, the entry will begin after the ship has changed speed, and that speed is constant for 20 minutes.

Note that each change in direction or change in speed is referred to as an "Event". These will appear as dots on the Mapping System. Further note that the journey between two "Events" is known as a "Path Step". This will appear as segments, between the dots, on the mapping system.

In addition, the crew can look at the raw data if desired. Each new line of raw data will be recorded follows: Date and Time; XYZ Direction; and Speed. For example:

- Date and Time: Oct 15, 2132 at 14:30
- XYZ Direction: XY = 118 degrees // Z = 41 degrees
- Speed: 200 km per hour

The computer will then have a long list of raw data, for the directions and speed of the travel. This raw data will be stored in the main data storage of the ship's computer. The crew can access this list at any time. However, the real use of the raw data will be to create final calculations and maps.

Automatic Calculations Every 8 Hours

The next step in the process is to calculate the total time, and total distance, for each of the previous entries in the list of raw data. This will be handled by a separate program, and automatically updated every 8 hours. This set of data will be used later to create the final calculations and maps.

Note that is will be updated every 8 hours. We do want regular updates, though we do not need to update continuously. Therefore, automatic updates every 8 hours is a reasonable period.

Mapping System: Tell Computer to Create Map for Each Step

Remember we said at the beginning that we can trace our path, as if painting a line behind us wherever we go. Therefore, the final operation available to us is drawing the path of our travels on the computer screen.

The first map is the Step Map. This map is based on the series of steps taken during the travel. This is the list from the second computer program. The user simply calls up the history of events, then selects the "show map" button. The entire set of individual path steps are now shown on the screen. The user is then able to see the entire history of the journey, with each step.

Additional options will allow the user to highlight only the changes in speed or where speed is maintained as constant. The user can also zoom closer on any region of the map. These options allow the user to get a visual understanding of the information. The map can also be rotated in various ways, to allow the 3-D reality to be understood on the screen.

Direct Path Map: Tell Computer to Create Direct Path Map

The final map we can create is the Direct Path Map. This is the map of the Direct Line, from Home Planet to current location.

This begins when the user begins the calculations for the Beacon Location. The user tells the computer to perform the final calculations. This operation will use the data from all the individual Path Steps, to calculate the Direct Angle and the Total Distance.

These values are then presented on the screen clearly. These values are also the official location of the Beacon on all beacon maps.

The numbers of the precise location are most important. (These values are for the Direct Angle and Direct Distance). Yet the Direct Path Map provides a useful visual aid.

When the user selects the "Create Direct Path Map" option, the computer will make two graphs. The first graph is the series of step by step Path Steps. Then the computer will make a second graph showing the Direct Path. This map is best created in a 3-D program, where the user can see all dimensions, and rotate the path drawing as desired.

The important data will be listed on a side column. This includes the Direct Angle and the Total Distance.

Total Review of the Path Drawing Program

The Path Drawing Program is the essential tool for knowing the precise location of each Space Beacon we place along the path. The "Program" consists of multiple parts, including the sophisticated mechanical engineering, the measurement devices, the data collection, and the computer analysis. The final result is the precise knowledge of the Direct Angle and Direct Distance between the Earth and the Beacon.

The process begins with the sophisticated mechanical engineering of the ship. The ship needs to be able to adjust to the exact direction and speed as desired by the crew. Similarly, the devices of the ship need to be able to know the exact XYZ degrees the ship is traveling, and the exact speed the ship is traveling, at any time.

This information is sent to the crew as feedback. This information is also sent to computer for storage of this data. The computer then creates a list of "events", where each event is a change in direction or a change in speed. This can be read as a long list of travel events, each with a time date stamp, followed by direction and speed.

A second computer accesses this data every 8 hours, to perform calculations and make its own list of events. This computer will analyze each two events, then determine the total time between each event. Using the total time, combined with the speed, the computer will determine the distance traveled, in that direction, during that time.

This information can also be viewed by the crew. They can also view as maps, where each line segment represents a travel path on the journey. However, the most important use of this information is determining the exact location of the new Beacon.

When the ship stops at its next location, for placing the next beacon, the crew will instruct the third computer program to perform the final calculations. This program will analyze all the path events from the second program, to create the Overall Path. Specifically, this program will determine the Direct XYZ angle from Earth to the current location. The program will also determine the exact Distance, along the Direct Path.

These results are used as the official location markers for the Space Beacon. Specifically, each Space Beacon will be located by the Direct XYZ Angle from Earth, and the Exact Distance between Earth and this location, along this direct line. The results can also be drawn on a 3-D map, as a visual aid. This is the process of the Path Drawing Program.

Chapter 4:

The Star-Map Room

Star Map Room: Overview

The "Star Map Room" is an immersive experience for viewing stars at any location. The room is a dome shape, where the stars are displayed on the entire dome. It then becomes as if you are standing in that location.

The stars shown are actual photographs, taken from ships on previous travels. Thus, the user can choose a location in space, and the computer will select the photos taken from that location. What you see then are actual photos taken from that location.

The display itself is a collection of smaller Display Screens. Each Display Screen is given a specific photo, taken at a specific angle. The entire Dome then shows the entire region of photos, which is the entire region of stars.

Because these are digital photos, the user can perform many choices, to get the exact Star-Map he desires. The user simply choses a location, any location where photos have been taken, then the computer will gather the appropriate photos, and send each to the appropriate screens. You are now immersed in the actual photos of the stars, as if you are there.

Furthermore, because these are digital displays, the user can call up data on specific stars. This data will be displayed on a screen next to the star of interest. The user can also ask the computer to display nearby Space Beacons and Supply Containers.

The entire Star-Map System is designed as a navigational aid for ships traveling through space. These Star-Maps can also be built on any planet.

Star Map Room: User Perspective

The Star-Map Room is an immersive mapping system. The room has a dome display, where stars are shown on the screen around you. Thus, when you stand in the room, it is as if you are looking at the stars which surround the ship.

The Dome Display works as a series of screens, similar to digital display screens. There are numerous most sized screens (rather than one big screen), each approximately 2 feet per side. Each will display its own image. Then, taken together, the Dome Display will show all of the stars, in all the positions.

The Star-Map Room will also have a model ship on a pedestal, which is connected to the real ship. Thus, as the real ship moves, so does the replica ship. Then, as the replica ship pivots, the images on each screen will change. When these many display screens are viewed together, as the full Dome Display, it will be as if the stars change in position corresponding to the ship's direction.

Taking Pictures of the Stars and Other Objects

The images will begin as live feed, from cameras on the top and bottom of the ship. However, these images will also be captured as screen shots, and stored in the computer. These stored images will be uploaded to future ships, and used in the Star-Map Room in various ways.

In addition, specific pictures are taken at pre-determined distances. The operators will point the cameras in specific angles, and label the pictures according to strict file naming. Thus, at specified distances, the crew will take quality pictures, of each viewing angle around the ship.

These photos will be uploaded to each Star-Map Database, on every ship and planet. Over time, the database becomes more comprehensive. Each next ship, on the next journey, will have a Star-Map Room which can show more regions of space, and with greater detail.

Note that there is a very specific protocol for how these pictures will be taken. The cameras will be located on specific regions of the outer ship, each one able to point in specific angles (X and Z angles). The details for the protocol, and the positions of the cameras, are described in the publication: "Databases and Mapping Systems in the Space Corps".

Main Viewing Options in the Star-Map Room

This immersive mapping system can be used for a variety of purposes. Specifically, there are 3 Viewing Options:

1. Live Image View
2. Recorded Still Image View
3. Remote Location View

Live Image View

In the "Live Image View", there are cameras on the ship which provide a direct image feed to the screen display. What you see in the Star-Map Room is exactly what exists around you in space.

Recorded Still Image View

The "Recorded Still Image View" is a collection of image stills taken from previous voyages. These images will match your exact location, and be refreshed every hour. This is useful for detailed study of the stars, because you can use the computer to select stars for closer view or for more information. You will then be accessing the computer data storage, yet can superimpose factual information on the screen of the stars.

Remote Location View

The "Remote Location View" will allow you to choose other locations in space, far away from your current location, and see what the stars will look like from that location. These images are again taken from still images recorded during previous journeys. Yet now you can see ahead.

You can choose two or more reference stars, tell the computer which part of the dome to display those stars, and the mapping system will fill in the rest. You will therefore get an immersive preview of what the stars will look like, when you get to that position in space.

Star Map Room:
Overview Design Drawing

In the drawing below, we see the basic design of the Star-Map Room. The outermost square (A) is the Square Outer Room. This has the general structure of a normal room. The Dome Structure (B) is built within the regular square room. The interior of the Dome Structure is covered with individual Display Screens (C). Each Display Screen will show its own small section of stars; the entire Dome Display, of all the screens, will become a realistic immersive experience within the region of space.

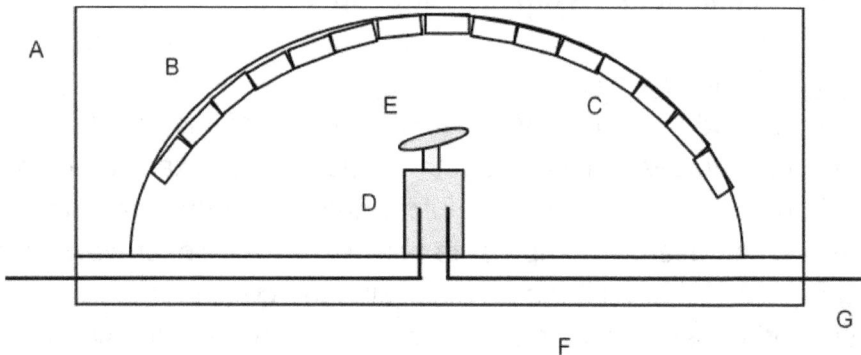

A = Outer Square Room

B = Dome Structure

C = Display Screens

D = Center Pedestal

E = Replica Ship, on Pivot

F = Wiring Layer, between floors

G = Cables Connecting Computer to Display and Other Systems

In the center of the room is the Center Pedestal (D). This Pedestal houses the main computer for the Star-Map. The cables which transmit data are also connected to the computer inside the Pedestal. These wires are laid in a Wiring Layer (F) beneath the Map Room, then extend in various directions to the screens (and to other computer systems). The top of the Pedestal has the Replica Ship (E), which matches course direction of the real ship. Each of these components are described below.

Chapter 5:

Design Drawings for the Star-Map Room

Star Map Room: Dome Structure

The Dome of the Star-Map Room is a very important structure. Therefore we must understand the physical structure and the arrangement of electronics for the Dome.

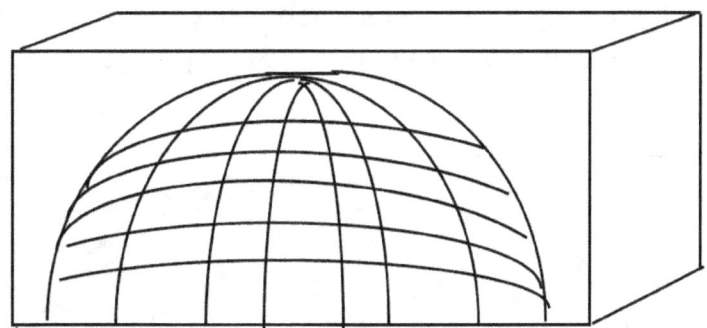

The Dome Structure is contained within the Outer Square Room. This is very important, as the Dome is *not* there to be a physical structure. Rather, the Dome is to be a sophisticated visual display. Therefore the Outer Square Room provides all the physical structure for our Map Room. The Dome is then placed within the Square Room.

The Dome is obviously spherical in X-Y directions. In some of our drawings only one cross-section is shown, yet what we show in the one cross-section is essentially duplicated 360 degrees.

The Dome Structure itself is made from material which is light in weight, yet very strong. Note that the individual Display Screens will fit into the spaces between the grids of the dome, similar to glass in a window pane. Thus, the material will not only support the dome structure, but will also hold the computer screens in place.

The Dome Structure will extend from the floor of the Square Room to just below the ceiling. However, the Display Screens will begin only 3 feet from the ground.

There Map-Room is entered through two consecutive doorways. The first doorway opens to the square room. The second doorway enters the dome structure. These are not shown, but are assumed in design.

Star Map Room:
Screen Size and Placement

The immersive experience of the Star-Map Room is based on the Dome Display. This Dome Display is actually made of numerous small screens, each showing a different section of the stars. Taken together, this will simulate the reality of stars as seen from any selected location.

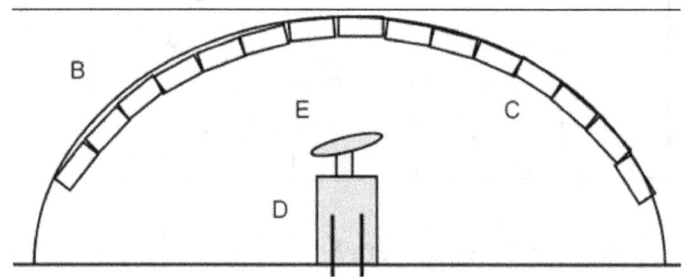

Each flat screen is known as a "Display Screen", and will display its own photo of stars. The entire set of Display Screens is the "Dome Display". The "Dome Display" is the immersive view, from the combination of the individual "Display Screens".

Each Display Screen will be placed between the grids of the Dome Structure, similar to glass in a window pane. Thus, each screen will be inserted and secured to the grid sections. In the drawings, the screens are exaggerated to better understand the design. Yet in reality these are flat-screen components, which sit within the grid sections.

Each Display Screen will be approximately 2 feet per side. The exact geometry and dimensions of each Screen will depend on the dimensions of the spaces between grid sections.

However, we desire the smaller screens because this will be better for physical stability (of both the dome and the screens). Each space between the dome support lines will be small, making the dome structure more stable; this also means that the screens are not likely to bend or fall from their locations.

Star Map Room:
Dome Support Lines and Data Cables

In addition, there will be cables traveling through some of the dome supports. Thus, many of the grid lines will be hollow, allowing cables to travel through these paths. The grid sections will continue to provide strength, yet the interior is hollow for data cables and electrical wiring.

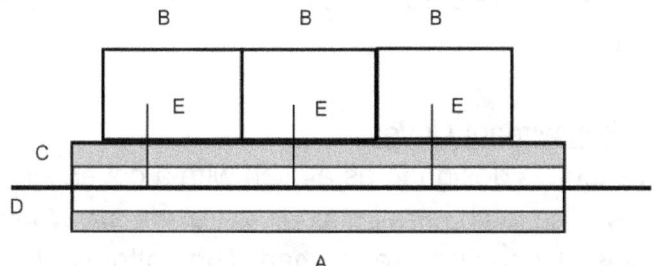

A = Dome Structure Grid Support Line

B = Display Screen

C = Structural Support Micro Layers

D = Data Cable or Power Line

E = Individual Data Wires, to each Display Screen

The Dome Structure exists due to the numerous Structural Support Lines, likely made of a plastic or alloy. The Display Screens are inserted between the Support Lines, as in a window pane. The interior of each line is hollow, which allows cables to be laid inside. We can lay two types of cables: data cables and electrical power cables. From the cable, there are individual wires which lead into each of the Data Screens.

Thus: the interior of the Structural Support Lines can carry the cables for electrical power and data transfer (star photos) to each Display Screen. Yet the surrounding areas of the Structural Lines contain neoprene or other material. This provides structural support for the Line, as well as insulation for the wiring.

In this way, the data cables for star photos, and the electrical power needed, for each Screen are lined along the same routes as the grid structure itself. It is efficient in many ways.

Star Map Room:
X-Axis Structural Lines and Data Cables

The Dome Structure exists because of a grid of Structural Lines. These are essentially poles, made of strong yet light-weight materials. These Structural Lines are built in generally two directions: X-Axis and Y-Axis.

The X-Axis Structural Lines will not only provide structure, but will also contain the wiring needed for the Display Screens. Thus, each X-Axis Line will provide wiring for the Display Screens above. The details are explained here.

X-Axis Line: Ditch without Cover

The X-Axis Line is designed as a ditch with a cover. An enlarged view is shown below. The ditch remains open when we are working with the wiring, then closed when we are finished. The bottom of the ditch is a strong material, which provides the strength of the X-Line. The data cables are laid in the ditch, on top of the strength material. Later, when sealed, this ditch becomes a hollow conduit with wires inside.

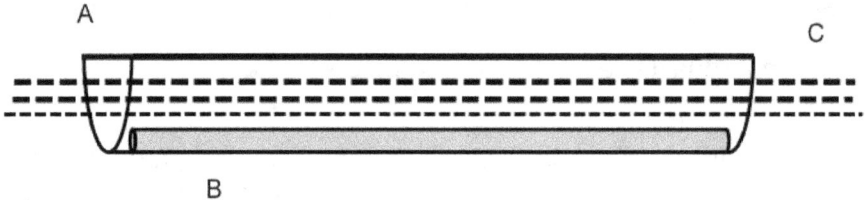

X-Axis Structural Support
and Cable Conduit

Open Ditch

A = Open Ditch Design

B = Structural Support Material at Bottom

C = Data Cables and Electrical Wires Laid Across

X-Axis Line: Segments and Full Length

Each X-Axis Line has two functions. The first is to be the X-Axis Structural Supports. The second is to be Conduits for the Data Cables and power lines. The X-Lines will likely be built in sections. The engineers will decide best what lengths to use for each section. For example each section of X-Axis Line could be approximately equal to 5 or 10 Display Screen Lengths. Each section will be designed as the Open Ditch Design.

The Data Cables and Electrical Power Lines will then be laid in the ditch. Note that there will be one data cable for each Display Screen. Also notice that each data cable will extend to different lengths; this is because each Display Screen will be different distances from the computer. Thus, we will lay several distinct data wires along the ditch, and each data cable will reach just far enough to reach the next Display Screen in the row.

This means the ends of each data cable will be increased approximately 4 feet steps. This is because each Display Screen Space will be approximately every 3 feet, plus another few inches for the depth of the Screen Tray. The data cable for each Display Screen will then reach from the X-Axis Conduit, to the Tray, and to the Display. (This can be seen in the next diagram).

Regarding the Electrical Power, this line will be different. We need only one line for all Display Screens on the row. There is no difference in the type of electricity for the Displays, therefore we use simple branches. There is one electrical cable, across the entire X-Axis Conduit. Then individual branches of wire extend from the main power line, again approximately every 4 feet, to reach each Display Screen. This is shown I the next diagram.

Note that each of these wires, for data and for power, can be quite thin. For the Data Cable, we are only showing one photo at a time, on each Display Screen. At times we may overlay the Space Beacons, or replace the photo with the data for a specific star. Yet that is all. Therefore, the amount of data through the data wire is relatively small.

Similarly, the amount of electrical power usage will be small. The Display Screens will show the stars. Of course, the Display will be bright enough so that the viewer can see the stars. Yet the amount of power required per Display will not be that much, and the total power line per X-Axis Conduit will be modest side.

The practical result is this: though we have numerous wires in the conduit, for many displays on the row, these wires are easily laid together in the conduit of each X-Axis Line.

X-Axis Line: Cover with Notches for Exit Wires

Each of the Display Screens require two wires. The first is the data cable, which provides the star photos and data. The second is the electrical power. There are distinct wires for the data, and distinct branches of a main line for the electrical power. These two wires are connected to the corresponding Display Screen.

When these wires are connected, the Conduit is sealed with a top cover. The cover is closed with a simple clasp. There are also two notches in the top, so that as the top is placed on conduit, the notches surround the wires, and the wires poke through the closure. This is illustrated below.

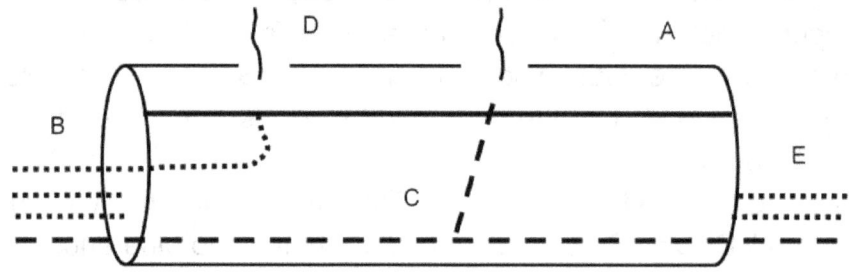

X-Axis Structural Support
and Cable Conduit

Closed Conduit

A = Closed Cover for X-Axis Conduit

B = Data Line, for this specific Display Screen

C = Electrical Power, Main Line, with Branch to Display Screen

D = Notches in Cover, to surround Data and Electrical Wires

E = Other Data Cables, to the next Display Screens

Underground Layer for Initial Cable Conduits

For the sake of being complete, we should remember that the first Data Cables are laid underground, in the wiring room layer. This is a crawl space, under the Star-Map Room. The Data Cables come down from the central computer in the Pedestal, then to the Layer underneath the Room.

There, we find several Conduits, spread in several directions. Each of these conduits contains several of the individual Data Cables, all of which are going to Display Screens in the same general region of the Dome.

Thus, these underground conduits bring many of individual data cables to a region of the Dome. From there, the conduit divides into smaller conduits, each going to different Rows, yet within that same general region. It is each of these smaller conduits, in each region of the dome, that is lined along one of the X-Axis Conduits, as illustrated above.

Y-Axis Lines for Structure Only; X-Axis Lines for Dual Purpose

Note that the Y-Axis Structural Lines will be pure structural Lines. These will be solid structural supports. They are not hollow. As such, the Y-Axis Lines will provide full structure for the Dome.

These Y-Axis Lines will also hold the Trays where the Display Screens are sitting. The Tray Hooks attach to the Y-Axis Lines, then the Display Screens sit inside the Trays. (See later for diagrams on the Trays)

Then X-Axis Line of the Row provides two purposes. As illustrated above, the X-Axis Line is actually a conduit. The exterior and the bottom lining of the X-Axis Conduit provide support structure for the Dome. The wires within the conduit will provide the Star Photos, Electrical Power, and other Data for the Displays.

Star Map Room:
The Display-Screen Trays

The Display Screens will be inserted into the Grid Space using the Screen Trays. The Screen Tray is the object that will be attached to the Dome Structure, then the Display Screen is set against the rims of the Screen Tray. Thus, the Screen Tray holds the Display Screen in place.

Note that the entire process is done from the back side of the Dome Structure. The Screen Tray is inserted into the Grid Space, and fixed to the Dome Structural Support Lines. The Screen Tray has a wide rim, on all four sides, capable of supporting the Display Screen. Thus, the Screen Tray is physically attached to the Dome Structure, while the Display Screens rest gently against the rims of the Tray.

The wiring is then attached. As described earlier, there are electrical power wires and data wires which have been laid along some of the grid structure lines. These wires are then attached to the back of the Display Screen. The Display Screen is then fully installed.

The Screen Tray

The "Screen Tray" is designed to be the actual component which holds the Display Screen in place. The Screen Tray has two important functions. The first function is attach to the Structure of the Dome. The second function is be able to support the Display Screen in place, where the Display faces the viewer.

The general structure of the Tray is rectangle. The Tray has two types of extensions. The first is a set of outward extensions, which are known as "Tray Hooks". These Tray Hooks will sit on the Grid Support Lines. In addition, the Tray will be permanently fixed to Grid Support Lines at these hook locations, using small screws or bolts.

The other type of extension is an inward extension. This is where the Display Screen actually sits. Specifically, the Tray extends an inch deep, and the bottom of this tray is an open square. This bottom has "rims" on which the Display Screen will sit, and then project the stars through the open square.

Display-Screen Tray:
General Side View

The drawing below shows the general structure of the Tray, as viewed from the side. The Tray Hooks (B) sit on the Dome Structure Lines (A). The two are then fastened together at the top and sides, at (C).

The Tray is approximately 1 inch deep. (The distance is exaggerated here to show clarity of design). The bottom is an open rectangle, with Support Rims. These Rims are approximately 1 inch inches wide, around the entire rectangle. The Display Screen will then sit on the Support Rims.

Therefore the Display Screen will be structurally supported, regardless of the angle in the Dome. The Display Screen will be encased on 3 sides, supported by the Rims, and held in place with the Tray being attached to the Dome Structure.

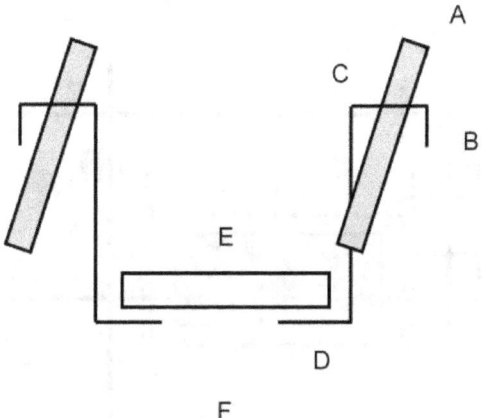

A = Dome Structure Lines

B = Tray Hook

C = Attachment Location (small screws)

D = Display Support Rims

E = Display Screen

F = Stars Shown to Human Viewer

Display-Screen Tray:
Dome Structure and Depth Perspective

The following diagram shows the Tray for the Display Screen, as placed in one of the spaces of the Dome Structure. The picture also shows something of the depth perspective of the Tray.

This view is looking at the Tray and Dome Structure from behind the dome. This is the view you will have when inserting the Tray and the Display Screen.

Note that the Tray Hooks sit on the Y axis structural lines, while the data cables are routed through the X axis structural lines. (The data cable and power wires are then connected to the Display, sitting in the Tray).

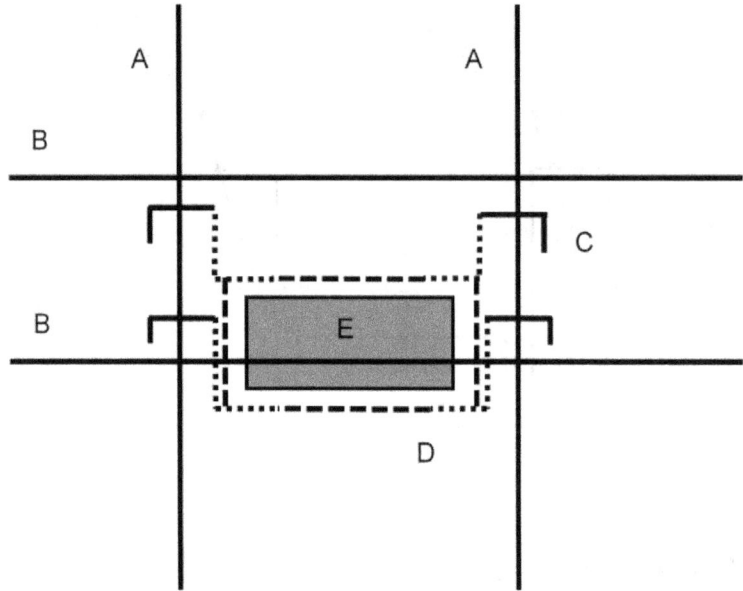

A = Dome Structural Lines: Y direction

B = Dome Structural Lines: X Direction

C = Tray Hooks

D = Tray for Display Screen

E = Display Screen, back side, sitting on Support Rims

Two Types of Trays:
Placed Every Other Position

There are two types of Trays, and these will be placed every other position in the Dome. Both types are essentially the same, except for the top extension. Type "A" has a Hook, whereas Type B has a longer extension. The basic concept is this: the extension on Type B will lay on top of the extension of Type A. The two will then be fastened together, and ideally both will then be fastened to the Structural Line underneath.

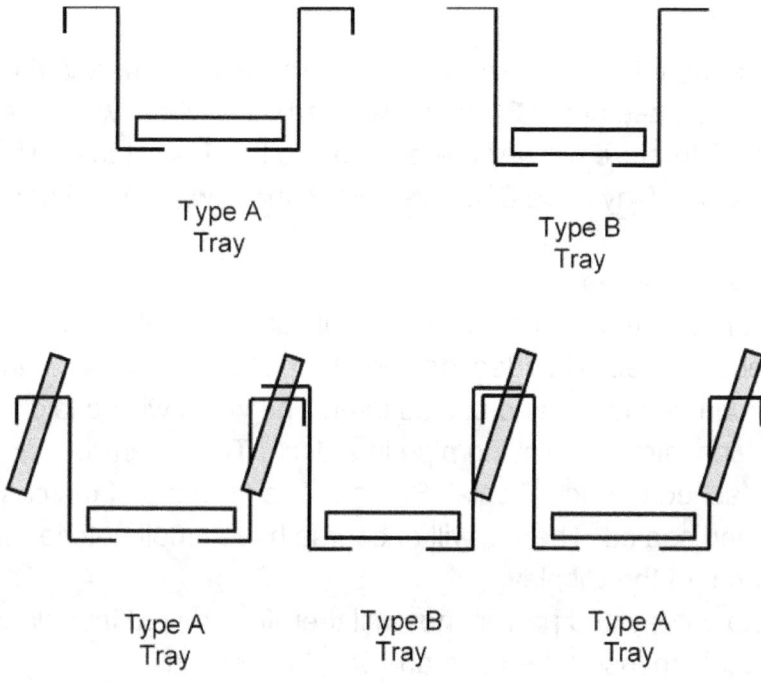

Type A
Tray

Type B
Tray

Type A
Tray

Type B
Tray

Type A
Tray

Also, as a technical note, the sides of Type B tray will be slightly taller. This will allow the extensions of Type B to sit on the Hooks of Type A.

The "Easy-Go System"
for Installing Display Screens

The Easy-Go System

This installation process will be referred to as the "Easy-Go" System. This is because each Tray and Display Screen is "Easy" to install; and then we can quickly "Go" to the next Display Screen.

The process is listed as steps here. Refer to the illustrations on the previous pages.

1. Select the Row

We will attach the Trays and Display Screens one row at a time. Note also that the X Lines of the Structure will carry the wires, while the Y Lines of the Dome Structure will be where we place the Tray Hooks. Therefore, we will place the Trays and Displays across the row, one row at a time.

2. Lay the Data Cables

The X Lines of the Structural Dome will carry the data cables and electrical power lines. The ideal design of the X Line is to have an open top, which can be closed and sealed later. The wires will be laid inside the Structural Line, similar to laying pipe in a ditch. The extension wires will soon be attached to each Display Screen. Then the top of the casing can be closed and sealed. The top will of course have small notches which fit over the wires to the Display.

The data cables and power lines will therefore be put into place at this time. We will then insert the trays and Display Screens.

3. Lay the Screen Trays, and Attach: Three at a Time

Next lay the Screen Tray into the Space. The Tray Hooks will sit on the Y axis Dome Structure Lines. The Tray will then extend an inch or so beyond. Remember we are doing this from the back side of the dome, and therefore the tray will extend (a small amount) to the interior of the dome.

We attach the trays to the Y-axis Line using small screws, at top and side of each Hook. Note that it is best to lay these trays three at a time. This is because the center tray will be Tray Style B, which connects the three trays together. Thus, for Tray B, the screw goes through extensions of B, then A, then into the Y-Axis Line.

4. Attach Wiring to Display Screen

Each Display Screen will require two wire connections. The first is the data cable, which sends photos from the computer to the Display Screen. The second is the electrical power cable, which provides power for each device. These two wires extend from the main line (in the X-Axis Ditch). Thus, we simply attach these wires to respective ports. The Display is now connected.

5. Note on Wiring Lines in the X-Axis

Note that the Main Cable Line in the X-Axis Ditch, is actually a set of individual wires. Each Display Screen will require its own wire. This is because each Display Screen will be getting its own star photos. Thus, we will have a series of data wires running along each X-Axis Line.

Each Data Wire will end at a different location; this end is the location of the next Space, where we will insert the Display Screen. Thus, the number of data wires becomes fewer after each Display Connection.

The Electrical Power Wire is slightly different. This will be one cable all the way through, with branch wires at each Space. This is sufficient, as the electricity being sent to each Display is the same.

6. Gently Place Display Screen into Tray

We are now ready to place the Display Screen into the Tray. Note that it is better to attach the wires to the Display Screen prior to this, as the Display is above the Tray, in your hands, and easier to work with. Thus, we gently set the Display Screen into the Tray, where it will remain.

If we have three Trays installed at the same time, as suggested above, then proceed to the next Display Screens for those Trays. Attach the wires, and gently set into the Tray.

7. Place the Cover on the X-Axis Ditch

Having finished with the wires at this location, the wires can be covered and protected. This cover will also provide additional structural support and insulation.

The Cover for the X-Line exists in segments; each segment is placed between the Y-Axis Lines. The Cover also has two notches which slide over the wires which lead to the Display. The cover is then fastened with a simple snap or hook.

Note that the Display Screen has no back cover. This is to provide air circulation for each Display. Also, the design of the tray is enough to hold the Display in place without backing.

8. Proceed and Repeat

We are now ready to move to the next Display Screen Space. We can install the next set of Trays, connect the next Display Screens, and set them into place. We cover the X-Axis Segments.

This process is repeated, along the entire row of Spaces, around the Dome. We then repeat the process for each successive row.

*Using this process, it is easy to install the Trays and the Display Screens. Easy to install, then proceed to the next set. The enter set of Display Screens will be installed without any complications.

Star Map Room:
Center Pedestal and Wiring

In the center of the room is the Center Pedestal (D). This Pedestal will house the computer system for the Star-Map. The Data Cables (G) extend from the computer to the Display Screens and other computers.

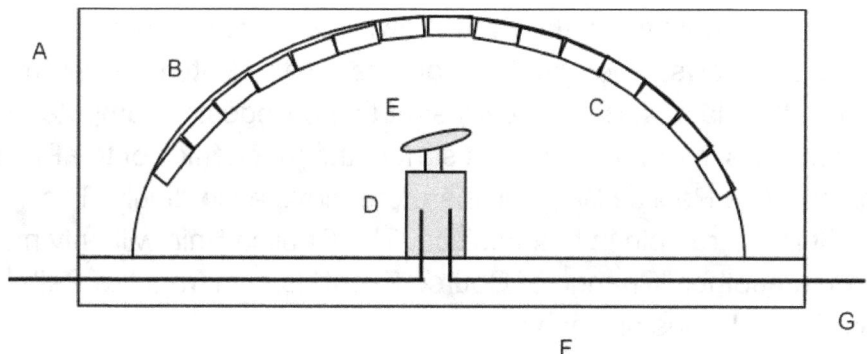

There is a Wiring Layer under the floor (F). This is essentially a crawl space, under the entire Map Room. This layer allows many data cables to be spread out in multiple directions. Several of these data cables are connected to other computer systems, including the Path Drawing Program, and the Cameras which take photos of the star regions.

Other cables will lead to the Display Screens. Specifically, the cables will spread out in several directions under the floor to the dome walls, then into the Dome Structure Lines (as shown earlier), and eventually to the specific Display Screens.

Note also that there is a space between the Dome Structure and the walls of the Square Room. This space allows for additional wiring needed, including the electrical power and some data wiring to each screen.

Star Map Room:
Center Pedestal and Replica Ship

On top of the Pedestal is the Replica Ship (E), which pivots to match the course of the real ship. This Replica Ship sits atop a pivot, and will move in both X and Z directions. The angle of the ship will match the angle of the real ship, relative to the Earth.

This data is provided by the Path Drawing Program. Specifically, the process is as follows: when the first database of the Path Program records an "Event", this data is automatically sent to the Pedestal Computer in the Map Room. This information is then sent to the mechanics of the Pivot Mechanism. The Replica Ship will therefore move accordingly. The Replica Ship will remain in that position. The Replica Ship will only move again when another "Change of Course Event" is sent from the Path Program to the Pedestal Computer.

Note also that the orientation of the Replica Ship is independent of any other systems. Therefore, the Replica Ship will always point in the course direction of the actual ship, relative to the 0-0 angles of Earth, at all times. This is true regardless of what stars are being shown on the Dome Display, or what Viewing Option is used.

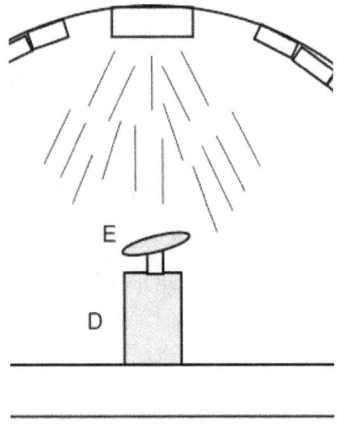

Star Map Room:
Map Room Lighting

The Dome Structure will require an overhead light. This lighting system will be placed in the ceiling of the Dome, in the direct center. The light will then come down from above, and spread over the Center Pedestal. Thus, when you walk in the door, and turn on the light switch, you will see the Center Pedestal fully bathed in the light from above.

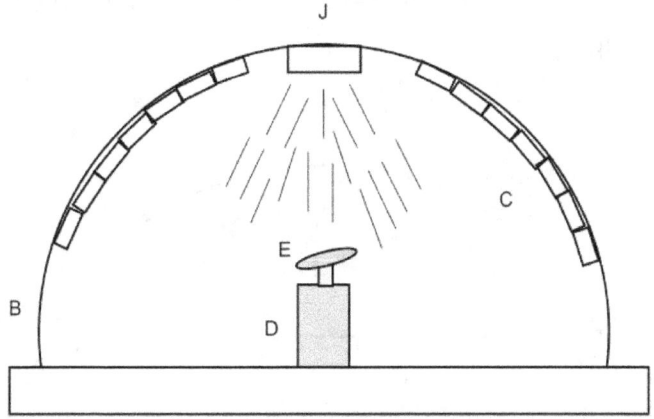

This light from above will illuminate the Central Pedestal, as you work in the Star-Map Room. This light will be sufficient for the user to operate the computer at the Pedestal, choosing specific Viewing Options and specific stars. The user will simply turn off the lights when ready to view the entire Star-Map in the Dome.

    B = Dome Structure

    C = Display Screens

    D = Center Pedestal

    E = Replica Ship

    J = Overhead Lighting System

Note that the light will be wired from the wall, as in any normal lighting system. There is a switch by the door, then the wiring extends through one of the dome structure-lines, to the lighting system on top. Note also that we need only one overhead light. This is all that is required.

Star Map Room:
Screens for Remote Location View

When using the Remote Location View, there are four specific Display Screens. These are labeled "A, B, C, D" in the diagram below. These are also labeled in the computer options in the Map Room.

We select two stars from the database, and two of the four Display Screens. The computer then figures out which photos to gather that match this viewing perspective. The system will then fills out the entire Dome Display accordingly.

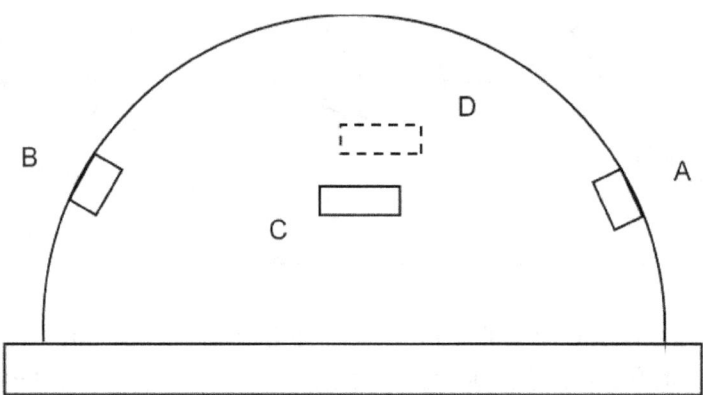

Specifically: when we select the Remote Location View, the computer will prompt us with "Screen A". We then enter the name of the star of interest to show on Screen A. The computer prompts us for "Screen B", which is directly opposite Screen A. We select a reference star (or planet) for this screen. (We can also select options for C and D if desired).

The computer will then use those stars, at those screens, to collect all the photos which were taken from this viewing perspective. The computer will then send specific star photos to each of the Display Screens as appropriate, to create the full immersive display.

This creates the complete immersion experience for the user, within the Star-Map, as if standing in any remote location.

Chapter 6:

Space Beacon as Navigational Guide

Overview of Space Beacons

When we travel into deep space, we will need a series of Space Beacons. These Space Beacons will provide two essential services:

1. Location Guides for Navigation
2. Communication Relay Stations

1. Location Guides for Navigation

As location guides, the Space Beacons will emit specific frequencies of electromagnetic energy, at a specific pulse rate. This signal will be detected by any passing ship, who will then match the signal with a list of signals from all Space Beacons. These signals will therefore act similarly to lighthouses, and confirm the location of the ship in space.

Note that most of the Space Beacons will remain stationary. These Beacons will be placed in deep space, far from other objects. Thus, there will be no gravitational interaction with other objects, or collisions with objects. Thus, the Space Beacons will remain essentially in permanent locations. This is important, as the Beacons will remain as absolute location markers. Therefore, these Beacons as location markers will be used throughout the galaxy, and beyond.

2. Communication Relay Stations

As Communication Relay Stations, the Space Beacons will transmit signals effectively in deep space. Using a series of these Space Beacons, distant planets can communicate with each other, regardless of the vast distance. This system will be used across the galaxy.

EM signals are well known to spread out as they travel, and therefore the signals reaching the detector becomes weaker with distance. This can be a problem for communication in deep space. Therefore, we use a series of Communication Relay Stations, each of which will capture the signal, boost the intensity, then re-emit.

Using this series of relay stations, the signal which reaches the destination will be much stronger. This is important when communicating with our fellow humans over such long distances.

The design and operation details of the Space Beacon Network are fully presented in many other publications by the authors. In the sections below, we will focus on Beacons as Location Guides.

Space Beacon as Location Guidance System

The first essential function of the Space Beacon will be as location guides throughout space. The Space Beacons will indeed be an essential navigation system for space exploration.

Navigation Using Stars

Remember that navigation requires knowing your location at all times. However, this requires having some fixed locations as reference points. When traveling in deep space, this can be more difficult.

Within a solar system, we can use other celestial objects such as moons and planets. However, when traveling in deep space, there will be fewer such reference objects.

Of course, we will use the many distant stars as reference points. These stars are so far away that they are essentially fixed in location. Therefore, we can use these stars to guide our position.

This is the basic method of navigation used for centuries. However, there are many limitations to this method. It is easy to deviate from the straight course, while thinking you are heading correctly. This becomes especially significant when traveling across deep space. Therefore, other methods should be used for navigation.

Navigation Using Beacons

The Space Beacons will provide absolute navigation points in space. Each Space Beacon will act as a lighthouse and location marker. As a lighthouse, the beacon emits a signal. The signal is unique, which is then correlated to a specific beacon, and then displayed on a Beacon Map. The crew will thus know their location.

Furthermore, because the Beacon is in deep space, the location of the Beacon is essentially permanent. Therefore, we can be certain that each Space Beacon is indeed at that exact location in space. Therefore, all passing ships will be able to confirm or adjust their location, when passing near the next Space Beacon.

EM Emission and Unique Beacon Signal

Note that the EM emission is very important for the Beacon Signal. The signal itself is a burst of EM energy. Therefore the emission process and intensity of the signal are important designs.

Furthermore, the "unique" signal is based on several factors. These factors include: frequency of the EM burst; amount of time the burst is on; and amount of time the burst is off. These three factors, combined, make the signal unique. This is how we correlate the Beacon Signal to a specific Space Beacon, and thus can display on the Beacon Map.

Therefore, because of the importance of EM emission in the Beacon Signal (and thus as Navigational Guide), we will devote several sections of this publication on the details of the Beacon Signal.

Signal Emitting Process

The general process of the Space Beacon Signal is simple. An internal device will create photon systems of specific frequency. A second device will emit these signals at specific on/off burst intervals.

There are four emitting dishes, one on each side of the post. Therefore, these specific frequencies are emitted in all directions with each burst.

Any space ship in the area will be able to receive these signals. The ship will know which Beacon is sending the signals, and their distance from the Beacon. This information will be correlated to the Beacon Map. The ship's location is then displayed for the crew.

This system is therefore used as confirmation of ship's location, or letting the crew know how to alter their course if necessary.

Identification of Beacon #1: EM Frequency

There are two factors which will specifically identify each Space Beacon. The first is EM Frequency, the second is the Emission Rate of signals being sent. Both have comparable uses in everyday experiences. The details are discussed below.

EM Frequency as Beacon Identification

The first identifier is the EM Frequency. We use this method already in radio communications. All FM stations, airplanes, ocean ships, Wi-Fi systems, and other communication systems, use a specific bandwidth of EM energy. Furthermore, each radio station (and many specific vehicles) is given a specific frequency for communication. We will use a similar system for the Space Beacons.

All Space Beacons will be given a specific bandwidth of EM frequencies. In this way, the receiving devices of the ships can be created to be tuned to that range of EM, for detecting the Beacon Signals.

Note however, that the EM frequencies used for Beacon Signals must be very different from the signals used for the voice and data communication systems. This is important to prevent interference. It is also important for the purposes of knowing we are getting the identification signal of the Beacon, rather than the human messages.

Regarding the *specific* Space Beacon, each space beacon will be given its own unique EM frequency. No other Space Beacon will use that same frequency.

Thus: all Space Beacons emit their ID signal within the same range, allowing use of radio detection equipment designed for that range (rather than the whole EM spectrum); yet each specific Space Beacon will emit its own specific EM frequency. The detector will be able to register the frequency of the Space Beacon, and the computer system will match to the database of Space Beacons.

Identification of Beacon #2: On/Off Burst Rate

The second factor used for the identification of the Space Beacon is the On/Off Burst Timing of the signal. This is similar to the lighthouse and the pulsar. That is, the rate of EM Bursts sent will identify the location.

For the lighthouse, the light rotates at a specific rate. The ocean ship will see the light appear, disappear, and appear again. The rate of rotation of the light, emitted from the lighthouse, is one method used to identify one lighthouse from another.

Thus, the crew of the ocean ship observes the frequency of light and dark, which then identifies a specific lighthouse. The crew will then know exactly where they are. By knowing the location and rate of light rotation of each lighthouse, the crew will be able to know their exact location.

We will use a similar system for our Space Beacons. As with the lighthouse, we will have a series of "light/dark" cycle times. The amount of time of the "light" vs the "dark" will identify the specific Space Beacon, just as is done for identification of the specific lighthouse.

There are a few differences however. In the Space Beacon, we use EM frequencies other than visible light. Therefore the signal will not be visible to the crew. However, the ship's sensors will detect these frequencies. We also use sequence of "emit on/emit off" rather than any type of rotation. Yet the result is essentially the same.

Note that the "Emission On" Time should be a significant amount of time. This is important for the passing ship to be able to detect the signal. It is also important that the crew knows that the signal came from the Beacon, rather than some random space event.

Always remember that each Space Beacon will be given its own rate of on/off, as the second part of the Beacon Identification factor.

Unique Signal As Combination of Factors

As we develop our Beacon Network throughout space, we may find ourselves duplicating some factors. However, no space beacon will have all factors the same. Thus, some Space Beacons, very far apart, may have the same EM frequency, but not the same cycle rate. Other space beacons will have the same cycle rate, but not the same frequency. Yet no beacon will have the combination of the same frequency and cycle rate.

Thus, using all three factors, each Space Beacon will emit a unique signal. The Beacon Databases will record this information, to ensure no duplication.

Intensity of Emission

The "intensity" of the Beacon Signal has practical implications. With greater intensity, the signal will reach ships further away. However, this requires more power and more photon supply, which will then need to be replenished more often.

The intensity can also be used to calculate distance of the ship from the beacon. The decrease in intensity with distance is a simple formula. Therefore, knowing the initial intensity of the signal to the received intensity, we can determine the distance of the ship from the beacon.

Intensity of Emission: Basic Science

First, let us remember what "intensity" really means. The "intensity" of an EM signal is in fact the number of photons in the initial cluster. When photons of the same EM energy are emitted at the same moment, they group into a cluster. It is this cluster that travels through space.

However, these photons gradually break apart. They become smaller clusters, breaking off and traveling in a regular cone shape. Therefore, the intensity of the signal received is a smaller group of photons.

Therefore: in general terms, the Intensity of an EM signal is the number of photons, with the same EM energy, which travel in a cluster. The initial intensity is the number of photons emitted from the source at any one moment. However, the intensity of the signal received is a smaller group of photons; a smaller piece of the initial larger grouping.

Understanding this science of EM Intensity is important when designing the Space Beacon systems.

Intensity of Emission: What Initial Intensity to Use

The Intensity of Emission is another factor to consider. It is well known that all EM emissions will spread out during travel, and therefore each section of the cone becomes weaker. The signal which reaches the detector will therefore depend on three things:

1. The initial Intensity of the emission

2. The proximity of the passing ship

3. The sensitivity of the detecting equipment.

When we emit with a greater intensity, there are more initial photons in the cluster. This means the signal can travel much further, and the pieces which break off will be easier to detect at these far distances.

However, this also requires greater internal resources for each emission. This will consume power of the Beacons more quickly.

Furthermore, there will be many times when no ship is nearby, and the signal is not being detected. Therefore, we must evaluate how far we really need to send the Beacon Signal, as a strong signal, compared to the rate of power usage in the Beacon.

One factor which will help is knowing that the Space Beacons will be placed along a path. The initial placement will be done by space craft heading to a specific moon or planet. The ships which come later, traveling to that same moon or planet, will also follow this same path.

Therefore, we can think of these Space Beacons as similar to signs on a road. Most of the Space Beacons will be placed along very specific paths, and all ships which travel these paths will usually travel the same distance from the Space Beacons. With this knowledge, we can better plan for the Intensity of the initial signal from the Space Beacon.

Note that it is important to have the same intensity for all Space Beacons. This is because the distance from ship to beacon can be calculated from the amount of diminished signal. Knowing the initial intensity, and measuring the received intensity, we can determine the distance of the ship from the Beacon. This will help the crews with more precise navigation.

See other publications on this topic for additional details.

Using Space Beacon as Navigational Guide:
General Review

The primary use of the Space Beacons is to be a Navigational Guide for space travelers. The Space Beacons will serve as location guides and mileage markers. Using these Beacons, the ship's crew will know their exact location, and make course adjustments as needed.

Therefore, we should review the practical steps involved in using the Space Beacons as a navigational guide. Notice that we now will be viewing this from the perspective of the ship's crew, rather than from the Beacon itself.

Basic Process of Beacon Signal as Navigational Guide

The Space Beacon emits a unique signal. This a stream of EM energy, sent in regular series of bursts. Thus, the Space Beacon is very similar to a lighthouse. Furthermore, the specific signal of the Beacon Signal is associated with a specific beacon. This allows the crew to match the signal to a beacon, then to the Beacon Map. Therefore, as the ship will know its own location as compared to the Beacon Map.

Additional important details are discussed below.

1. Beacon Emits Unique ID Signal

The process begins with the Space Beacon. Each Beacon emits a specific signal, as a stream of EM energy. There are several factors which make this signal unique: EM Frequency; Time of Stream On; Time of Stream Off; and Initial Intensity. The Beacon emits its unique signal, in all four directions, in its regular on/off cycle, without end.

2. Expecting the Signal when Near the Region

The ship will expect to receive the signal when near the region. The crew will know they are in the region of the next Space Beacon using several other navigational aids. These tools include the Star-Map Room, the Beacon Map, and the Path Drawing Program.

These are each mapping systems, some of which include actual star photos of the region (taken by previous ships). Therefore, the crew will compare their visual reality with the star photos to determine when they are getting close to the next Beacon.

3. Ship Receives Beacon ID Signal

The ship has receivers which are designed specifically for receiving the Beacon Signals. When the ship gets within range, the signal is detected. The Signal is then analyzed, the data is sent to a computer.

Thus, the crew can look at the computer screen for the Beacon Signal Analyzer, which will then display each of the factors. These are of course, the frequency, timing cycle, and intensity received.

4. Computer and Crew Determine Beacon Number

The computer will then automatically search through the Beacon Location Database, and match the Signal Factors to those in the Database. The Beacon Number will then be determined.

This process can be made simpler by instructing the computer to only search for the Beacons on a specific Beacon Line. This setting will remain in place, until manually changed.

Note that the Beacon Signal is being regularly received. Therefore the data page will be continuously renewed. The one factor that will change will be the received intensity; as the ship gets closer to the Beacon the received intensity will increase.

The human crew should also verify the data, and verify that the Beacon Number is the Beacon they expected.

5. Beacon Location Determined on Beacon Map

The Space Beacon can now be determined on the Beacon Map. The crew will open the Beacon Map Program, then select the specific Space Beacon. The Mapping Program will then display the Space Beacon, with a real photo of the stars behind the Beacon.

The user can then perform a variety of other requests, for different views, including expanded distance views and rotation.

The crew can now be certain that they are in the area of the Space Beacon, and therefore use the Mapping Program to see where they are in relation to other Beacons, their position along the Beacon Line, and other comparative locations.

6. Star Photos Compared to Visual Sightings, for Angle of Approach

The crew can then pull up the star photos, to determine their angle of approach. This is best done in the Star-Map Room.

Thus, the crew goes to the Star-Map Room and selects this Space Beacon as the primary object. The Dome Display then shows all the stars relative to the Space Beacon. The crew will then compare this Dome Display to their visual reality of the stars outside the ship. From this, they will know their angle of approach relative to the Beacon.

This will help the Ship to maintain an exact parallel path along the Beacon Line. If the ship has angled away from the Beacon Line, this will be noticed when comparing the visual reality of stars to the Star-Map Room. The crew can adjust accordingly.

Thus, each time the ship passes another Space Beacon, they will perfect their course to be more exactly parallel to the Beacon Line. They will attempt to maintain this course between Beacons.

7. Intensity Received will Determine Distance from Beacon

The crew will also monitor the received intensity. This can be used to know the distance between the ship and the Beacon. The signal spreads in regular fashion, in a cone shape. Therefore, the distance between the ship and beacon can be calculated. Note that this is the side distance, between the path line of the ship, and the Beacon Line.

The initial intensity from each Beacon is known, and in the database. Thus, the computer compares the received intensity to the known initial intensity, and determines the distance between emitter and receiver. There is of course some variation due the angle of the ship versus the Beacon, which is why we adjust for the ship being parallel first.

If the ship needs to be closer to the Beacon Line, the crew will use this process to move closer, and to the desired distance. This can be especially useful for ships which have wandered from the Beacon Line to explore, and wish to return to the Beacon Line again.

8. Crew Determines Exact Location of Ship

The crew is now able to determine its exact position in space. The crew knows the Beacon Signal and Beacon ID, and thus places on the Beacon Map. The visual reality is compared to the star photos. The intensity has been used to know the distance from the Beacon Line. Thus, from this information, the crew knows their exact location in space.

9. Ship Maintains Course or Adjusts Course

The ship will then use this information, and the various mapping systems, to maintain its course. The crew will of course adjust their course as needed, using the Beacon Signal and Background Stars as guides.

10. Send Message to Home Planet of Ship Location

As the ship passes the Space Beacon, the crew will send a message to the home planet. This message tells the home planet that the ship has reached this specific beacon. Thus, the Beacon serves as a mileage marker. The home planet will know that the ship is on course, and has reached this location in space safely.

The ship will also do the "Casual Inspection" of the Beacon, as discussed earlier, and message the report.

The crew will also tell the home planet the expected date to reach the next beacon. If no message is received, the home planet will send a message to check the crew's safety, and send a rescue ship if needed.

11. Ship Continues to Next Space Beacon

The ship will then continue onward to the next Space Beacon. The crew has confirmed their location. The crew has perfected their course, based on the Beacon Signal and Star Maps. Thus, the ship should be able to maintain an essentially parallel course, to the next Beacon.

Chapter 7:

Beacon Placement
and Knowing Official Location

Overview of Knowing Beacon Location

The purpose of the Space Beacon as Location Guidance is to help the ships traveling in deep space to better know their locations. This requires several important location steps.

We must first know the exact locations of the Beacons when we place them. Each Space Beacon must also have a unique identifying signal which can be associated with each specific Beacon. These will be placed on comprehensive map of Space Beacons.

Each Beacon will then emit the signals, in all four directions. When a ship passes the Beacon, the ship will receive the signal, and know the location of that Beacon. The crew can also note the intensity of the signal, relative to the known original intensity, as well as the direction.

Taken together, this information can be used by the navigation crew to determine their exact location. This will be a significant benefit for all ships traveling in deep space, where there are few other location references.

Some of the practical details regarding Location of the Beacon are described in the next sections. In addition, there is a separate section on the practical points of Placement of the Beacon, which is closer to the end of this publication.

Placement of Beacon and Selecting Location

All Space Beacons will be placed by the initial ship which travels in that direction. Each space ship will have multiple Space Beacons in the storage rooms. The crew will be trained to setup these Space Beacons and test them before leaving the area.

The general locations will be determined by the Communications Division of the Corps. However, the specific location will be determined by the crew of the first Space Craft in the area.

The crew will pause the ship in each of the approximate locations determined by the Corps. Once the Space Beacon is in place and operational, the crew will determine the exact location of the Space Beacon, relative to Earth and other celestial objects.

The location of each Space Beacon will be measured with reference to Earth. Thus, each Space Beacon will be measured as a precise distance and direction angle from Earth. In addition, secondary distances to nearest stars will also be listed, as approximations.

This information will be added to the computer database of the ship, into the and into their mapping programs. The information is also then sent to the Corps back at Earth, where this database and map of Beacons will be regularly updated.

Knowing Specific Location During Placement: The Path Drawing Program

The crew will then use a series of navigational tools and computer programs to determine the precise location of the Space Beacon they are placing. The primary tool for knowing the location of the Beacon during placement will be a computer program which tracks the exact path taken by the space craft. As the space craft travels, the navigation equipment will be able to measure exact distance traveled, in each direction the ship travels. (This can be in any of the spherical directions).

This information will be stored in a computer program, and drawn on a series of computer screens. Therefore, from this information, the ship crew will know exactly where they are, with reference to Earth. The crew will then use that information to determine the exact distance and direction, in a straight line, from Earth to the Beacon.

This set of location coordinates will be the "exact location" of that Space Beacon. This information is then sent to the Corps, and is placed on all maps used by the Corps. All space ships will have the most updated maps of Beacon Locations before embarking.

Ship Perspective:
Using the Beacons for Navigation

Once the Space Beacons are set up, these Space Beacons will operate for years. They will be powered by nuclear energy. Their photon supply for emissions will come from the cosmic particles and background radiation. Thus, the Space Beacons will be able to operate for years.

Any space craft which travels along this path, to the destination moon or planet, can use these Beacons as a location guide.

The ship has a series of detectors which are specifically designed to register the EM emissions of the Beacon Frequencies. When a Beacon signal is detected, the crew is notified. At this point, both human and automated systems perform the same investigations. (Having both human minds and computers examining will serve as double checks).

First, the Identification Factors are measured against the database of known Space Beacons. Thus, the specific Space Beacon is identified.

Second, the mapping system is brought out, with the locations of the Space Beacons. The crew will then know exactly where this Beacons is located, relative to Earth and other celestial objects.

The navigation crew (and automated system) will then note the direction of the signal compared to their ship. This lets them know the direction of the Beacon relative to themselves.

Finally, the intensity of the signal is compared to the known intensity of the emitting signal. Simple calculations allow the navigation crew to know the distance, based on resulting amount of diminishing intensity.

Using all of these factors together, the navigation crew, and the automated system, can place the location of the ship at a very precise location relative to the Space Beacon; and therefore know their location relative to Earth and other space objects.

*Note that many of the modern space ships will also have the path drawing programs described above. These will be primary navigation tools for all ships. However, the accuracy of the path drawing program in the first ship will be the most advanced and most accurate possible.

Other path drawing programs in other ships may vary in complexity, and thus mechanical parts, and costs. Therefore, the navigation crew of any space ship (after the first in the area) will be able to use a combination of their path drawing program and the Space Beacons to guide the crew.

Beacon Signal Detectors on Side and Front of Ship

For the Main Beacon, we emit the Beacon Signal in four directions. Two of these directions face the passing ships. These are the main Beacon Signal Directions which the passing ships will receive.

The Beacon will also emit the signal in two other directions, which are facing the planets, and therefore essentially face the oncoming ships. The ship will be able to receive some of this Beacon Signal (as cone shape spread) while nearing the Beacon. This is likely the first signal to be received by the ship. Yet both signals will be received and analyzed as navigational guides.

Note that the ship will use two different receivers, one facing the side, the other facing forward. This is necessary to detect each of the Beacon Signals (front emission and side emission) being emitted from any Beacon. The computer will know which receiver is obtaining the signal, then take this into account in the mapping program.

The computer will also be able to receive Beacon Navigation Signals from both the front of the ship and the side of the ship, at the same time. The computer will then analyze each signal, and use computations involving data from both signals. This will more accurately determine the location of the ship, relative to the Beacon.

Determining Exact Location Using Space Beacon

Therefore, using a combination of specific EM frequency emitted, and the rate of on/off of the emission, each Space Beacon can be specifically identified by any passing ship.

The crew will be able to match the specific identification signal to the database and map of Space Beacons. This will determine (or confirm) the location of the ship. The ship will be able to know its exact location from the identification of the Space Beacon.

The crew will also be able to make more precise knowledge of their location by noting the direction where the signal came from, and the diminished intensity of the received signal.

Furthermore, if the ship is between two Beacons and is able to receive signals from both, the ship can use both signals to determine their location. This can be very valuable in deep space, which is mostly empty.

Chapter 8:

Beacon Databases

Beacon Databases: Types and Purposes

There are several databases related to the Space Beacons. Some of these databases exist for the design and maintenance of the Space Beacon. Other databases exist for the purposes of the Space Beacon as Navigational Guide.

The main Beacon Databases we will discuss are the following:

- Beacon Design Database
- Beacon Maintenance Database
- Beacon Location Database

Furthermore, the specific data to collect for the Beacon Location Database is very important for navigational purposes. Therefore we will spend several pages discussing this data.

See Other Publication for Additional Details

Note also that a separate publication was written, specifically for Databases and Maps. Please see "Databases and Mapping Systems in the Space Corps" for complete details on all databases.

Beacon Design Database

The Beacon Design Database is the engineering design database for each of the space beacons. This database will be used to catalog each of the Space Beacons, and list many of the important components. Official drawings will also be connected to the listings.

The primary users of the Beacon Design Database will be the engineering designers, manufacturers, and the Corps Headquarters. This information will also be used by the Beacon Maintenance Crew, to understand the technical construction of any Beacon they are inspecting.

Database Factors in the Beacon Design Database

The "Beacon Design Database" should include the following data:

- Beacon Identification Number
- Beacon Category Type (Main, Node, Solar System)
- Design Type
- Energy Supply Type
- Estimated Date of Replacing Energy Supply
- Message Recording System (Yes/No)
- Message Storage Box Type
- Estimated Replacement Date of Message Box
- Beacon ID Signal: Frequency Range Options
- Beacon ID Signal: Timing Burst Options
- Beacon ID Signal: Intensity Range Options
- Manufacturing Company
- Lead Design Engineer
- Lead Manufacturing Engineer
- Date Completed
- Links to Associated Engineering Drawings
- Other Design Factors as desired

Each of these data categories are explained briefly below. Note also that each of these factors are described in greater detail in other publications.

Beacon Identification Number

The Beacon Identification Number is the unique number which is assigned to that Beacon. This is the number for which all beacons, in all of the databases, will be identified. This number will be etched into the surface of the space beacon, in a very visible location.

The Beacon Category Type is the purpose of the Beacon. There are only a few Types. The beacons of "Main Space Beacon"; "Node Beacon"; and "Solar System Beacon" are the most common. Additional categories include the Stations, such as "Node Station" and "Solar System Station".

The Beacon Category should be listed in two columns, as the phrase, and as a Code Letter. For example: "Main Space Beacon / A", and "Node Beacon / B". The Code Letters are useful for short-hand entries.

The Design Type is the basic design of the Space Beacon. This is similar to the model of a car. The names should be simple phrases, such as "Hamilton", "Mavery", and "Pindle". (Specific naming choices are left to the design team, yet should always be simple one-word names).

Most of the Space Beacons (which are used for the same purpose) will use the same design type; or at least begin with that Design Type before modifications. Thus, the user of the database will get a quick understanding when reading the Design Type. This will also link to the drawings of the "standard" engineering of that Design Type.

Finally, note that the Official Beacon ID Number should be a combination of all these categories. For example: "Mavery-A-273". This tells us that the Space Beacon is of the Mavery Design; it is used as a Main Space Beacon (in a Beacon Line), and is Beacon #273 in the series.

Energy Type

The component which will need replacement most often will be the Energy Supply. Therefore, the Design Database should list the types of energy used, and the amount of energy contained in each Energy Supply. This will be followed by the estimated date when the Energy Supply needs to be replaced. The data should also indicate the number of Energy Pockets in the Space Beacon. This will be linked to drawings of how to remove and replace the Energy Supplies.

Message Storage Box

The Message Storage Box is optional, yet will need replacing. Therefore the Design Database should list whether or not the Space Beacon has the Message Storage Box, and the type of Storage Box which is used. This is followed by the estimated date of replacement.

Beacon ID Signal

Note that one of the important factors is the Beacon ID Signal. There are three separate categories which make the Beacon ID Signal:

- Frequency of Emitted EM Energy
- Timing of On/Off Burst (seconds flowing, seconds of nothing)
- Initial Intensity

It is important that no two beacons have the exact same Beacon ID Signal. Therefore, the Design Database and the Location Database will allow designers to see what Beacon Designs are being used, and therefore avoid anything which is too similar.

However, we also want to design the Beacons to be adjustable in their Beacon ID Signal. This allows us to change the settings of the Beacon ID Signal, without changing the internal wiring.

Therefore: the Design Team will designate a *range* of options, for each of the Signal ID categories. The range will be limited, yet adjustable. It is the range of options which is entered into the Design Database. Later, the exact values will be set by the Placement Crew, using a keypad on the beacon. These numbers are then entered into the official Location Database. Note also that the engineers, in collaboration with the Communications Division, will state the "standard" or "ideal" values for that Beacon. These are the values which the Placement Crew will enter, unless directed otherwise.

Exceptions to the Unique Beacon ID Signal

Each Space Beacon should have its own Beacon ID Signal. This is very important, for being able to match the specific beacon to a specific location on the Beacon Map.

However, there will be some exceptions. As we expand our Beacon Network throughout the galaxy, the total size of our Network will become quite vast. Thus, the ships which travel near one far corner of the network are not likely to travel near another far corner of the network. Thus, the ships are not likely to be confused.

Similarly, the signals from the beacon themselves will spread out across the miles. Therefore, with such vast distances, the signal from one beacon, in a very far corner, will become very weak near the center of the network, and almost undetectable by the location of the other far beacon.

This can be compared to radio stations, where there are many cities which have station that use the same frequency. When these stations are far enough apart, given their emitting intensities, then the signals will not interfere. We can use the same frequency, in different locations far apart.

This means we can, for beacons this far apart, use the same Beacon ID Signal. As our network expands, and we have settled the vast regions of the galaxy, we can use some of the same Beacon ID Signals. But again, only when the two beacons are so far apart that their signals will not be confused to any passing ship.

Associated Engineering Drawings in the Design Database

The Beacon Design Database will also have a collection of detailed drawings. The image program will obviously be a different program from the Database, yet they will be linked together. Thus, for many of the components listed in the Database, the user will be able to click on the word, to call up the image of that component.

The Associated Drawings should include: Beacon Design Type; Energy Supply Operation; Energy Supply Replacement; Relay System Details; and Beacon ID Signal Details.

Note that only the Design Database and the Maintenance Database will have these drawings, as they are not needed in the others.

Design Database: Physical Locations and Access

The Design Database will be physically located in the Communications Division, at the Corps Headquarters. However, all engineers associated with the design and construction of the space beacons, will have access to this database, through their secure internet. Any data entry and changes to data will be limited to specific engineering and communications staff.

Duplicate copies of the Design Database will be loaded to the computers of all cargo ships, primarily for the use of Maintenance Crews. The database will again be duplicated for the Communications Offices of each planet. Note that these versions are "read-only". No changes can be made to the information.

Beacon Maintenance Database

The "Beacon Maintenance Database" will record all of the maintenance activities, for each Space Beacon, throughout the network of beacons. This database will help to ensure that all Space Beacons are operating at optimum condition.

Note that this database will be organized differently, as there will be several different perspectives for viewing the data. We will therefore discuss the database primarily by viewing page. Within the discussions of each viewing page, we will discuss the details of the data within.

The Beacon Maintenance Database will be organized into the following Viewing Pages:

1. Quick View of Beacon Maintenance Status

2. Beacon ID Number View

3. Scheduled Maintenance View

Quick View of Beacon Maintenance Status

The "Quick View" of Beacon Maintenance is the first page of the database. The purpose of this View is to see the latest inspection, latest maintenance, and related information, for all Space Beacons. This will also be the first page, from which the user will reach all other Views.

The first two columns of this page are the Beacon ID Number and the Beacon Line. The rest of the columns are the essential data for this Viewing Page. The data columns on this page will include:

- Beacon ID Number
- Beacon Line Name
- Latest Inspection Date
- Type of Inspection: Distance, Manual, On-Board
- Latest Maintenance Date
- Ship Name of Inspection/Maintenance
- Team Leader of Inspection/Maintenance
- Notes from Recent Inspection/Maintenance
- Next Scheduled Maintenance Date

Remember that this is a Quick View, for all Space Beacons. The detailed information will be on other Viewing Pages.

Note that we can simplify the results by selecting the specific Beacon Line. By clicking on the phrase "Beacon Line" at the top, the computer will produce a pull-down menu of the Beacon Line options. This will shorten the Quick View Page to only those Beacons on that Beacon Line.

There will also be a search box, where the user can type the Beacon Line or Beacon Number, which will lead directly to that page.

Beacon ID Number Viewing Page

The details of inspection and maintenance activity, for each Space Beacon, will be on its own page. This page will list every activity, by date. Specifically, each row entry will be a date of inspection or maintenance activity. The column categories will be components and related information.

The Beacon ID Number Viewing Page can be accessed in one of several ways. In the Quick View Page, click the Category of Beacon ID Number. This will create a pull-down menu, from which you can select the Beacon ID. This can be simplified, by first limiting to the Beacon Line as described above. The Beacon Number can also be typed directly into the search box on the screen.

This Viewing Page is now for the one Beacon Only. The top of the page will show, for clarification, both the Beacon Line, and the Beacon ID Number. This will remind the user that which Beacon he is looking at. The Categories in the columns should include the following:

- Activity Date
- Inspection or Maintenance
- Ship Name
- Team Leader
- Link to Detailed Report
- Component Name (for rest of columns)

The rest of the columns will be specific systems and components. The results will be short responses, such as: "repaired", "replaced", "working normally", and so on. Anything not inspected or repaired will be left blank.

The short responses are important, as there will be many components. The user does not want to be overwhelmed when reading through all the components.

Detailed Maintenance Reports from Beacon ID Page

However, there will be detailed reports, written in free-form, for every inspection or maintenance. The Team Leader for the activity will write a few sentences or a few paragraphs, describing the activities. This will be in the computer as a pdf or similar document storage.

Therefore, the human to human exchange of detailed information on the maintenance activities can be accessed as needed, without cluttering the main Database Views.

Scheduled Maintenance Viewing Page

Another valuable Viewing Page is the Scheduled Maintenance Viewing Page. Many components will have scheduled maintenance.

Similar to the Quick View Page, the Scheduled Maintenance Viewing Page will list all Space Beacons. Each row will be a specific space beacon. Components which require scheduled maintenance are listed by column. Therefore, each cell data will be the scheduled date, for that component (column) for that beacon (row). This will allow any user to look at all of the Beacons, and all of the components of each Beacon, to assess when scheduled maintenance is coming, anywhere in the network.

This page can of course be narrowed to specific Beacon Line (similar to discussed above), and specific Beacon Number.

Note that the Quick View Page will show the next date for scheduled maintenance on each beacon, but not specifically what component. This View Page will therefore show the Scheduled Maintenance Dates, on all components, on all Beacons.

Beacon Design Drawings for Maintenance

The Beacon Maintenance Database must also have a separate data storage of each Beacon Design. The Beacon Maintenance Crews must have all the detailed diagrams, for each system and each component, of every Space Beacon and Satellite.

Thus, anyone of the Maintenance Crews can select the Beacon ID Number, and get a full set of detailed drawings. These files will also likely have specific diagrams on maintenance and repair, as well as reminders and best practices.

The Beacon Maintenance Crews will then be able to study these diagrams closely, before inspecting the beacon and operating on the beacon itself.

Sophisticated Search, to Create Reports

The Beacon Maintenance Database should also perform sophisticated searches. The database will have a Search Page. There will be many boxes of categories on the search page. For each option, the user will select his criteria, or leave blank. The program will then create the final result. This can be in the form of a spreadsheet, as lists on a document page, or a type of graph.

Note that the exact specifications of the Search Page will determined by those who will use the feature most often. This will also be a more advanced feature, used mostly by managers and planners at the Corps Headquarters, rather than the Maintenance Crew.

Physical Location, Access, and Updates of Database

The Master Database of the Maintenance Database will be at the Corps Headquarters. However, duplicates of the most current database will be loaded onto all cargo ships and maintenance ships.

Anyone who performs inspection or maintenance on the beacons will have authorization to make entries into the database. They will type all entries into their own database, on the ship.

When maintenance is completed, they will send their updates to the Corps Headquarters. This staff will then enter the updates into the Master Database of the Beacon Maintenance Database.

Thus: the Maintenance Beacon Database will exist at each planet, on each cargo ship, and each maintenance ship. However, the official version, the Master Database, will be located on Earth, at the Corps Headquarters. Each database will be updated as often as is realistically possible.

If there are any questions regarding the maintenance updates, send a message to the Corps on Earth, or contact with another maintenance crew nearby.

Beacon Location Database

The "Beacon Location Database" is the database which links to the Space Beacon Maps. Thus, the Beacon Location Database is the database which will be used as Navigation Guide. The Beacon Location Database is the database which will be used by most people, and used most often, when traveling through the galaxy.

The Beacon Location Database will therefore have the Official Location information. This includes the exact location of each Space Beacon, and the unique Beacon ID Signal. This information can then be presented on several Space Beacon Maps.

The Beacon Database is the first stage in this process. Therefore, in this section we will discuss the database itself. In the next chapter we will discuss the operations of the associated Mapping Programs.

Database Factors in the Beacon Location Database

The Beacon Location Database will have the categories, and specific values, which will create the Beacon Maps. Therefore, categories include the following:

- Beacon ID Number
- Beacon Line Name

- Starting Planet
- Direct Distance to Beacon
- Direct Angle to Beacon: X-Y Angle
- Direct Angle to Beacon: Z Angle

- Beacon ID Signal: EM Frequency
- Beacon ID Signal: Timing – Burst On
- Beacon ID Signal: Timing – Burst Off
- Beacon ID Signal: Initial Intensity

- Relay System: EM Frequencies

- Distance from Previous Beacon
- Brightest Star: Name, Angle, Distance (approximate)
- Photos of Stars

Understanding the Location Database Categories: Overview

The data for the Location Database can be grouped into general categories: Beacon ID; Location; Signal ID; and Relay System. The following is a brief explanation of each of the data categories listed above.

Remember also that each of these factors are described in greater detail in the next chapter, as well as in other publications.

Beacon Number

The Beacon Number is the official Beacon Number as described above. For example: "Mavery-A-273", tells us that the Space Beacon is of the Mavery Design; it is a Main Space Beacon (in a Beacon Line), and is Beacon #273 in the series. All factors are of the space beacon are associated to that Beacon using this number.

Beacon Line Name

The Beacon Line is the line of Space Beacons between two planets. Communications along the Beacon Line can be compared to passengers on rail line, traveling between two destinations. The Beacon Line name will be an important identifier for anyone who uses the Beacon Maps.

Starting Planet

The "Starting Planet" is the reference planet, from which all Beacon Locations are based. This is the planet where we designate zero distance and zero angle. Essentially, this reference planet is Earth.

However, this will not always be the case. As we expand our civilization throughout the galaxy, and develop or Beacon Network, there will be many Beacon Lines which have nothing to do with Earth. These are the Settlement to Settlement (S2S) Lines discussed earlier.

Therefore, the Starting Planet category exists so that everyone knows which planet we are starting from, especially when this planet is no longer Earth, but somewhere far away.

*For more details on using a different a Home Planet, see the chapter on Space Beacon Maps. The practical aspects of using multiple coordinate systems and viewing perspectives are discussed there.

Official Locations: Direct Distance and Direct Angle

The Official Location for each Space Beacon will be known by the Direct Distance and the Direct Angle. This will be Direct Distance between Earth and the Beacon; and the Direct Angle from the Earth to the Beacon.

The specifics of measuring these values have already been described in great detail. We will review these again in the next chapter.

Beacon ID Signal Data

Each Space Beacon will be identified by its Beacon ID Signal. This requires four specific data values: EM Frequency Emitted; Time Burst is On; Time Burst is Off; and the Initial Intensity.

The first three values together will determine the unique Beacon ID Signal, which can then be correlated to specific Beacon Number, and located on a Beacon Map. The fourth value, the initial intensity, will assist the crew in knowing their distance from that beacon.

Message Relay System Frequencies

The Relay System Frequencies are those frequencies used by the beacon for long-distance communication. These are the frequencies which the receiving dishes are designed to capture, and which the emitting dishes are designed to emit.

These values should, of course, be identical for all beacons on that Beacon Line. They should also be the same for all Beacon Lines. However, different communication frequencies can be used for different Beacon Lines, if desired.

Although this information is not technically needed for locating the beacon, it is useful for communications staff to have at their fingertips. This allows them to know which frequencies to use when sending messages through that Beacon Line, and which frequencies to use when wanting to receive messages from that Beacon Line. This is valuable for ships and planet based communications staff.

This information will also assist the Beacon Line Managers to ensure that the best frequencies are being used, for each of the operations of the Beacon Line.

Additional Useful Reference Points

The above values are the Official Values for the Beacon Location. However, the mapping programs and the navigation crews may find it useful to have other distance data.

These additional reference points include: 1) the distances between any two beacons; 2) the location of brightest star in a region to the beacon; and 3) photos of the surrounding stars.

The distances between space beacons is useful for the ships traveling this route, as a type of mileage marker. The Brightest Star will serve as a useful reference for the ships which have traveled away from the Beacon Line. This includes those ships which deliberately explore the unknown.

Photos of the surrounding stars will be associated with these data categories. This will allow the crew to know exactly which stars are being referenced as brightest stars, as well as to have a visual of what the crews should see in the area.

Physical Location and Access

The Beacon Location Database will be loaded onto the computers of every ship. This data will connect directly to the Beacon Mapping Programs, which will also be installed on every ship. This includes all passenger ships, all cargo ships, and all other types of ships.

The Master Database for the Beacon Location Database will be at the Communications Divisions in the Corps Headquarters. This is the only version of the database which can be altered. All other users will then get duplicates of the database.

The only exception for database modification is the ship with the Beacon Placement Crew. They will have their own version of this database, and be able to add essential data, as they put each new Beacon into position.

Additional Details on Beacon Location Data

In order for the Space Beacon Network to be effective as a Navigational System, we must have very accurate data. We require accurate data, for several categories, regarding each Beacon.

The details are important enough that we devote the entire next chapter to this topic. In addition, the processes of collecting this essential data are further elaborated in other publications, including: "Placement of Space Beacons".

Chapter 9:

Important Data for Beacon Location Database

General Overview of Data to Collect
for Beacon Location Database

All of the essential data, of each Space Beacon, must be entered into a master database. This data can then be used to create a set of Space Beacon Maps.

Thus, when a ship passes by any one Beacon, that ship will be able to know exactly which Beacon they are near, and then to confirm their location by comparing to the Beacon Maps.

However, in order for this to work as intended, we must first collect the essential data regarding each Space Beacon. Therefore, as soon as the Beacon has been put into place, the Measurement Crew will collect all of the essential data regarding this beacon.

The Measurement Crew will then enter that data into their own database, and begin to develop their own Beacon Maps. The crew will also send this information to the Corps Headquarters on Earth. This information will go into the Master Database of Beacons, as well as be the data for the Master Set of Beacon Maps.

Note also that when the crew sends this essential beacon data to the Corps, it will also be the first real usage of the beacon they just placed. This will be an effective test of the Beacon Relay System, at every location along the Beacon Line.

Categories of Data to Collect

There are many specific data facts to collect regarding each Space Beacon. The data can be grouped into general categories:

A. Beacon Equipment Identification

B. Exact Beacon Location: Distances and Angles

C. Additional Useful Location Measurements

D. Beacon Identification Signal

E. Background Star Photos

A) Beacon Equipment Identification

We begin with the Beacon Equipment Identification. This is simply the equipment ID number of that Beacon. All other data will be associated with this number. We will also include the name of the Beacon Line.

B) Exact Beacon Location

The Official Location of the Beacon, on all the Beacon Maps, will be given by the Exact Beacon Location Data. This includes the exact distance between Earth and the Beacon. This also includes the exact XYZ angle, from the equator of the Earth to the Space Beacon. Remember, these numbers are very important, and must be exact, because these are the official location measurements for the Beacon Location.

C) Additional Useful Location Measurements

In addition to the Official Location Measurements, we will also provide other useful location measurements. Note that these measurements are not official, and in some cases they not exact, yet they can be useful for other navigational purposes. These location measurements include the distances between beacons, approximate distances from each Beacon to brightest stars. This will also include distances to various planets.

D) Beacon Identification Signal

The Beacon Identification Signal is the burst of EM which is emitted by the Beacon. This is the signal which serves as a type of lighthouse or pulsar, and therefore serves as a navigational aid. The ID signal emitted by each Beacon is unique to that Beacon. No other Space Beacon in the network will have the exact same signal ID.

Therefore, the crew which sets the Beacon in place must be very knowledgeable regarding the Beacon Signal. They must not only know the signal for this Beacon, but must ensure that no two beacons have the same signal. (Each Beacon should have already been designed this way, yet the crews should also verify each signal is unique).

There are three factors to know regarding the Beacon ID signal. These are: 1) EM frequency, 2) the On/Off Timing of the Burst, and 3) Initial Intensity. The combination of 1 and 2 creates the unique ID signal for a Beacon. The initial Intensity is important for crews to estimate their distance from the Beacon.

E) Background Star Photos

The Background Star Photos are simply photos taken of the area, while the ship is parked near the Beacon. These photos will serve as visual aids for future crews passing the area. That is, the crews will be able to view these pictures, and compare to their actual travels. Regardless of their direction toward the Beacon, they will know what the arrangement of stars should look like, as the approach the Beacon, from any direction.

Specific List of Data to Collect

The complete list of data to collect is as follows. Each item will be further explained in a later section.

1. Beacon ID #

2. Beacon Line Name

3. Direct Angle from Earth to Beacon

4. Exact Distance from Earth to Beacon

5. Distance Between Adjacent Beacons

6. Distance to Final Planet (when known)

7. Distance to Nearest Star

8. Distance to Brightest Star

9. EM Frequency of Beacon ID Signal

10. On/Off Timing of Beacon Signal Bursts

11. Initial Intensity

12. Background Star Pictures

Each of these items will be explained in greater detail below. The data items are organized by Data Category.

Data Category A:
Beacon Equipment Identification

1. Beacon ID

The Beacon ID Number is the unique number associated with that Space Beacon. This number is assigned by the company manufacturing the Beacon. This number will be engraved on one of the main panels of the Beacon structure.

The manufacturing company will have all of these codes, and the design details for each beacon, in their database. Later, when this beacon is actually placed at the location, the crews will send a message to the Corps with this unique number. All subsequent information for that Beacon will be associated with that number.

1a) Beacon ID Number, Code System

In addition to the specific number, there will be an alpha code which designates the Category of Beacon, and the Design Type of the Beacon. This information will be useful for other purposes, including maintenance and design improvements. The Category Labels will be as follows:

A = Main Beacons (those along a Beacon Line)

B = Node Beacons

C = Solar System Beacons

D = Communication Satellites (orbiting the planets)

E = Photon Capture Beacons

F = Supply Containers (located in space, along the route)

The "Design Name" will of course be determined by the design team for that style of Beacon. The numbering will then be sequential, for each beacon of that design. For example: "A – Hannen – 238"

This will code will be translated as follows:
- A = Beacon type is Main Beacon
- Hannen = Design Model known as "Hannen"
- 238 = Sequential Number 238 for that design.

2. Beacon Line Name

The Beacon Line name is the second data fact to know. The computer systems will be able to list all Beacons on each Beacon Line. The maps will show each Beacon as a point along its Beacon Line. Therefore, it is important to state the Beacon Line name, for each Beacon Number.

The crew for any one ship is obviously placing each Beacon on the same Beacon Line. Yet for purposes of absolute clarification, the crew will state the name of the Beacon Line, along with the Beacon Number.

2a) Naming System for Beacon Lines

Note that there is a standard method for naming the Beacon Lines. This was explained earlier. A brief review is as follows:

There are two types of Beacon Lines, the Earth to Settlement Line, and the Settlement to Settlement Line. The first is abbreviated "E2S", while the second is abbreviated "S2S".

Notice that because there are two planets as end points of the line, both planets should be given as the name. The order of the name will always be alphabetical.

Thus, for example: the Beacon Line between Earth and the planet Dankhar will be stated as "E2S: Dankhar-Earth". This is how it will be written in the database, and on the Beacon Maps.

However, in conversational usage, the phrase "Earth" may be eliminated. If only one planet is mentioned, then it is assumed that the other planet is Earth.

For the Beacon Lines between two remote planets, the naming convention begins with S2S rather than E2S. The alpha order rule still applies. For example, a line of Space Beacons between the planets Dankhar and Shrivna will be listed as the "S2S Dankhar-Shrivna Line". However the line of Space Beacons between the planets Dankhar and Blanchette will be known as the "S2S Blanchette-Dankhar Line".

Data Category B:
Exact Beacon Location

The Official Beacon Location must be very precise. It is important that we know the exact location of the Space Beacon. This is the reason for the Path Drawing Program discussed earlier. Using this combination of mechanical engineering and computer analysis, the Measurement Crew will know, with precision, the location of the Space Beacon.

This exact location will be given with two values: the Direct Angle from Earth to Beacon, and the Direct Distance from Earth to Beacon. These values were explained earlier, and will be explained again below.

3. Direct Angle from Earth to Beacon

The Direct Angle from the Earth to the Beacon is taken from the equator of the Earth, when the Earth is at the position of orbit closest to the remote planet. When we draw a line from the Earth's equator at this location, directly to the Beacon, this line will be at an XYZ angle relative to the equator. This is the Direct Angle for that Space Beacon.

This Direct Angle will be determined using the Path Drawing Program, as described above. When the Beacon is placed into position, the Measurement Crew will perform the final calculations on the computer. This Direct Angle will be recorded as XY degree and Z degree, then messaged as one of the official location parameters.

4. Exact Distance from Earth to Beacon

The Direct Distance is very similar to the Direct Angle. This is the exact distance between Earth and the Space Beacon. As with the Direct Angle, the Earth will be at its orbit position where it is closest to the remote settlement. The distance is then measured from the equator of the Earth, at this orbit position, to the specific Space Beacon.

The exact measurement of the Direct Distance is very important. Therefore the distance will be measured and calculated using the Path Drawing Program described above.

4b. Units of Measurement for Distance

Note that the official unit of measurement for the Direct Distance to the Space Beacon will be in kilometers (km). This is necessary for the precision we desire. Also notice that the Path Drawing Program is capable of giving us this precision.

A secondary unit of "Astronomical Units" will also be given. This will allow us to compare this location to other approximate distances, such as to the nearest star. However, the Official Distance will be given in km, because this unit is most precise.

3c-4c Alternate Home Planets and Conversion Programs

Note that in the future, when the Beacon Line extends from another planet, the Distances and Angles will first be measured from the Home Planet. Yet these values will be converted to measurements from Earth. Thus, the values from Earth will always be the Official Values.

However, in the databases and maps, there will be an option to select a different Home Planet. All data will be converted, and maps adjusted accordingly.

Data Category C:
Additional Location Measurements

In addition to the Official Location Measurements discussed above, we will have other distance measurements related to the Space Beacon. These additional distance measurements will serve as useful guides, for a variety of purposes. These purposes include future explorations in other directions, away from the traditional Beacon Line Route.

Note however that some of these distances will be only approximate, because we have not traveled to those locations. These distances can only be estimates, and should be noted as such. Other distances will remain unknown during beacon placement, yet will become recorded and messaged when known.

5. Distance and Angle Between Adjacent Beacons

Each Space Beacon serves as a type of mileage marker. Therefore, the ship will often want to know the distance to the next Beacon. This will allow the crew to know where to expect the next Beacon.

For example, when the Beacon Signal is detected, the ship location is then confirmed. However, if the signal is not received when expected, then the ship is likely off course, and must make adjustments. Therefore, it is very valuable to know the exact distances from one Beacon to the next.

These values are precisely known during the Placement of each Space Beacon. Specifically: the Measurement Crew uses the Path Drawing Program to determine the Direct Angle and Direct Distance between the previous Beacon Location and the next Beacon Location. The crew then creates a set of tables, which lists the distances between each two Beacons along that Beacon Line.

6. Distance to Remote Planet

The Distance to the Remote Planet is also important to know. The initial distances will be approximate, given as Astronomical Units. When kilometers are listed, it should be stated that this distance is approximate.

However, as the ship travels the route from Earth to Remote Planet, the ship will be using its Path Drawing Program for very precise measurements. Therefore, when the ship reaches the remote planet, the exact distance between Earth and the Planet will be known. This precise measurement between the two planets will be recorded in the ship's database, and then messaged to the Corps Headquarters.

Notice that this distance will be given in kilometers. It will also be stated in the database that this value is "exact", rather than "approximate".

7. Distance to Nearest Star

The Measurement will estimate the distance between the Space Beacon and the nearest star. They can perform this estimate while parked at the Space Beacon. Note that because it is only an estimate, the distance can only be listed as Astronomical Units. (If km are used, it must be made clear that this is an estimate only).

The distance between Space Beacon and nearest star will have practical uses in the future.

The first application is being able to harness any photons and neutrinos from the star. Remember that we need to collect particles for use as beacon signal and for boosting the intensity of messages. Most of these supply particles will come from the background microwave photons in the area. However, if the star is close enough, then we may be able to capture some of the supply particles we need from that star. We may also use as partial solar power for operating the equipment.

Another application may be future Beacon Lines. Where there is a star, there is a solar system, with planets and moons. Some of these planets and moons may be inhabitable. Thus, we may be developing additional civilizations in those solar systems. We will be therefore be placing Beacon Lines along that route as well.

Thus, specifically: the Measurement Crew will give four data factors for the nearest star. These are: 1) name of the closest star, 2) noting it is the data for closest star, 3) XYZ angle of the nearest star, and 4) estimated distance of the nearest star.

8. Distance to Brightest Star

The Measurement crew will similarly be measuring the distance between the Space Beacon and the Brightest Star in the area.

As stated above for the closest star, the brightest star may be of potential use for harnessing the photons and neutrinos we need for our Beacon. Also, as stated above, the brightest star may have planets worth investigating. Ships may travel that route, and beacons may be placed along that route.

In addition, knowing the distance between each Space Beacon and brightest visible star in the area will be beneficial to navigation. The brightest stars will be seen for long distances, and therefore will give an approximate direction of the Space Beacons.

This will be particularly useful for the ships which explore regions of space beyond the Beacon Lines. Thus, as the ships travel into the deep space, in places without Beacons, they can estimate where the nearest Beacon Line will be. When they need to return to the Beacon Line, the brightest stars in the area will guide them back toward the Beacon Line.

Therefore, the ships can explore beyond the established routes, then return to the Beacon Line again, for effective communication, or as a guided path to the home planet. These brightest stars noted on the Beacon Map, with the associated Beacons on the map, will guide the ship back to the safety of the Beacon Line.

As with the data for the closest star to each Space Beacon, the Measurement Crew will give four data factors for the brightest star in the Beacon Region. These are: 1) name of the brightest star, 2) noting it is the data for brightest star, 3) XYZ angle of the brightest star, and 4) estimated distance of the brightest star.

This information will be available as options on the Beacon Maps. Using the data of the Space Beacons on any Beacon Line, and the Beacon's associated brightest stars, the crew of any ship can navigate away from the Beacon Line to explore further into space…and find their way back safely again.

Data Category D:
Beacon Identification Signal

The Beacon ID Signal is the specific signal emitted by each Space Beacon. Every signal is unique. Therefore, when a ship receives this signal, they will know exactly which Beacon it is. This can be matched to the Beacon Map, and confirm the ship's location.

The specific Beacon ID Signal is a combination of two factors: the frequency of EM emitted, and the timing of the On/Off bursts. Some beacons in the network will use the same frequency; and some will use the same timing; but no beacons will have the exact combination of the two factors. Thus, we can be assured that each signal is unique.

Note that the designed range options (for frequency and on/off cycle) will be recommended by the Corps Headquarters, for each Beacon. Their databases and oversight will ensure that each signal is unique.

The engineering team which manufactures the Beacon will then install a range of options accordingly. This is for both the frequency emitted and the cycle time. However, the final values will be entered on a keypad by the Placement Crew, when the Beacon is put into position.

Furthermore, reality of mechanics may differ from settings. Therefore the actual observation of the Beacon, in operation will be the official data for the Beacon Identification Signal.

9. EM Frequency of Beacon ID Signal

The first component of the unique Beacon ID Signal is the specific frequency of EM emitted. The Beacon will be designed with a range of frequencies (though a narrow range of frequencies). Then the specific frequency will be set by the Placement Crew.

When the Beacon is placed, and after the Beacon Signal has been operating for a few hours, the Measurement Crew will detect the exact frequency of the Beacon ID signal. This should of course be the same as the manual setting, though mechanics may alter the actual result.

Therefore the observed signal becomes the official EM frequency, of the Beacon ID Signal. This is the EM frequency which will be entered into the beacon databases and beacon maps.

9b. Installing EM Frequency Options of Beacon ID Signal

It is well known that Electromagnetic Energy exists in many frequencies. We can therefore choose which of these frequencies to use, in each of the Beacons.

There are in fact so many EM frequencies as options, that we could fill most of the galaxy with Beacons, each using a different EM frequency.

However, there are many practical factors to consider. This includes the fact that detection equipment can only be designed for specific ranges of frequencies. Therefore it makes most sense for our ships and beacons to have only the equipment needed to detect beacon signal frequencies of a specific range.

Furthermore, the Beacon ID Signal must not be the same range as the Communication Signals. Therefore it is most likely that certain ranges of the frequencies will be used, so as not to interfere with each other.

Thus, we have an almost unlimited range of frequencies to choose from as our Beacon Signals. Yet due to practical reasons, we will likely use a certain range of frequencies.

The Corps will determine the range of frequencies which make most scientific sense for use of the Beacon as Beacon ID signal. From this general range of frequencies, each Beacon will be given its own unique EM frequency.

This will first be designed into the beacon, as a much more narrow range of frequencies. Then the exact frequency is entered into the keypad of the beacon, by the Placement Crew during beacon placement.

10. On/Off Timing of Beacon Signal Bursts

The second factor in the Beacon ID Signal is the On/Off Timing of the burst. As discussed earlier, this is the amount of time that the EM burst will be emitted (on), and not emitted (off).

This timing of the burst, the On/Off, will act very much like a lighthouse or pulsar. The ship will therefore detect a repeated signal, every few seconds. The regular timing will confirm that the ship is indeed receiving a Beacon Signal. Furthermore, the exact amount of time, as On and as Off, will identify a specific beacon.

Therefore, the exact timing sequence of the Beacon ID Burst will be the second official data value to identify the specific Beacon.

10b. Amount of Time for Signal On and Off

The amount of time the burst is "on" should range anywhere from 10 to 60 seconds. Remember that it is important that the Beacon emits a continuous signal, for a period of time. This is to ensure that we can actually detect this as a signal, rather than just a few blips of photons.

The amount of time the burst is "off" can be any value. However, we suggest at least 3x the amount of time for off (no signal) compared to the time of on (steady signal). This is mostly to conserve energy supplies.

10c. Two Distinct Values, Either Can Be Adjusted

Note that there will be two times needed: the amount of time for the continuous burst (amount of time "on"), and the amount of time when there is no emission burst (amount of time "off"). Thus, the Official Data Values for the Cycle Time will have two values.

Notice also that either value can be adjusted to make a unique signal, as contrasted to other Beacons in the general region.

10d. Designing and Setting the On/Off Cycle

The On/Off Timing for each Beacon will be determined by the Communication Division at the Corps. This is to ensure that each Beacon has a unique signal. However, these values can also be adjusted by the Placement Crews. They will use a keypad near the Timing Mechanism, to make the final Timing Settings. These settings become the Official Data for the On/Off Timing Cycle.

Also remember that we require two values. The first is the amount of time of the continuous EM burst. The second is the amount of time between bursts, when there is no signal emission.

11. Initial Intensity

The Initial Intensity of the Space Beacon is the intensity of the EM burst emitted by the Space Beacon. Knowing this value will help the ships determine their distance from the Beacon.

In general, the initial intensity is the size of the cluster of photons emitted every second. This cluster will spread out as the photons travel. This spread is well known and creates the classic cone shape.

Therefore, using simple equations, the crew of any ship can measure their received intensity to the known initial intensity, and determine their distance from the Space Beacon.

11b. Designing and Setting the Initial Intensity

The intensity of the Beacon will be determined by engineers at the Corps, and built into the Beacon as the "standard intensity" for the Beacon. However, this intensity can be adjusted by future ships. As with other adjustments, a simple keypad near the Beacon Signal will allow workers to enter the exact value of the Initial Intensity. The result will be similar to making a light brighter or dimmer, as desired in the future.

Regarding the actual value of the Intensity to use, note that there is a balance to consider. Signals of greater intensity will be detected at greater distances. However, this will also require more photons and greater energy. In addition, ships may not be traveling that far away very often.

Therefore the process for setting the Initial Intensity will be as follows. The engineers at the Corps will determine the best range of intensities for each Space Beacon. The manufacturing company will then build the Space Beacon to be able to emit a signal within that range of intensities. The keypad will be attached, to set the Intensity by any space crew.

When the Placement Crew puts the Beacon in place, they will choose the "Standard Value" option on the keypad. This value is displayed, and will be the Official Value of Initial Intensity.

This value is the Intensity which will be recorded in all databases and mapping systems. This is the value of Initial Intensity which the ships will use to compare to the intensity of signal received, and thus calculate their distance from the Beacon.

Note that in the future, if the Corps decides to increase or decrease the intensity (within designed range), any crew can come by and make those changes using the keypad. They will then broadcast this message through the relay system.

Beacon ID Signal: Additional Practical Notes

The following are practical concepts, regarding the design and implementation of the Beacon Signal.

Beacon ID Signal: Manufactured and Coordinated

These two factors (photon frequency and on/off timing of bursts) should be determined by the Communications Division. This management will ensure that each beacon signal is unique.

The Communications Division will then work with the manufacturers of the Beacons. The engineers who design and bult the Beacons will install equipment which is specific to the parameters set by the Division. This includes installing Frequency Generators which will emit EM frequencies within the desired range. This also includes installing Cycle Timers which operate with the timing option ranges set forth by the Division.

At the same time, the engineers will design the equipment to be adjustable. There will be options for the Frequency. These options will be within a limited range, yet can be adjusted. This allows the frequency to be adjusted by space crews at any time.

Similarly, the Cycle Timers will be adjustable. The Division will designate the range of burst times (on) and no bursts (off). The Division will also set for the Ideal Values for each specific Beacon. Yet these values can be adjusted, within the engineered range, by any space crew in the future.

Design Database and Beacon Signal Oversight

The Communications Division of the Space Corps will keep a database of these factors, for all Space Beacons. This database will ensure that each Beacon Signal is unique.

These values will be found in the Beacon Design Database, which has factors related to design details for each Beacon. Note that this is a separate database from the database used for beacon maps.

Therefore, as new Space Beacons are being designed, the Corps will review the existing Beacon Signals. From this, they will assign the ranges of frequency and cycle timing for the next set of beacons.

Furthermore, from the database of Beacon Signals, the Corps will specify the ideal values for the Beacon Signal. This includes ideal values, for frequency, continuous burst time, and no-burst time. These values will then be engraved into the side panel of the beacon.

Adjusting Beacon ID Signal using Keypad

Using an electronic keypad, the crews can type in the value for the EM frequency to emitted; and again for the Timing of the On/Off Bursts. In this way, the Beacon ID Signal can be adjusted by the crews, while in space. Thus, our Beacon ID Signal can be both manufactured at the company, and adjustable by the crews in space.

The Initial Intensity can also be adjusted in this way. Using the keypad, any space crew can enter a new value for the Initial Intensity of the Signal.

However, again, the approximate parameters should be manufactured. This is to keep the combination of factors unique to each Beacon.

Adjusting Cycle Time vs Frequency

Regarding EM frequency, the options are limited. The equipment can only be designed to emit frequencies within a certain range. The greatest option for adjustability is for the On/Off Timing. This is the range which can be most widely adjusted. Therefore, if it the crew determines, for whatever reason, that the Beacon ID Signal needs to be adjusted, it is best to adjust the On/Off Timing, rather than the EM Frequency.

Observed Beacon ID Signal is Official Data

When the Beacon is actually being placed, the Beacon ID signal engineering should already be installed. However, engineering is rarely perfect. Therefore the crew will let the Beacon operate for several days. The crew will then determine, from observation, the exact EM Frequency, the exact timing of the On/Off bursts, and Initial Intensity.

*It is the actual observations of the Beacon ID Signal, in operation, that will be the official listing of the Beacon Signal in the databases and maps. These numbers should be very close to the designed values; yet the actual performance values will become the official data.

Accomplishing Multiple Important Objectives

Therefore, using the plans described above, we will accomplish multiple objectives for the Beacon Signal:

- the Frequency Range can be engineered into the Beacon
- the Frequency can be adjusted, with narrow range
- the Cycle Times are established with preferences
- the Cycle Times can be adjusted
- the Beacon Signal will be Unique, for each Space Beacon

135

Data Category E:
Background Star Photos

The final data to collect regarding the Space Beacon is a set of Background Star Pictures. These pictures will serve as visual reference for any ship traveling in the area. All future crews will know what the region should look like, near the next Beacon.

These pictures will also be used in the Star-Map Rooms. These pictures will be carefully labeled and loaded into the Star-Map Database ship, from which many Displays can be created.

12. Background Star Pictures

The Background Star Pictures will be very easy to create. The ship will have high quality cameras located on several regions of the ship. The crew will manually operate the camera, to take pictures of the stars in multiple directions.

The Measurement Crew will make sure that each photo looks good, to be of sufficient quality. The process may need to be repeated.

Each photo will be sent to Corps Headquarters, along with associated details. (See below). These photos will then be loaded into the Mapping Programs. Anyone can then look at the actual photos of the stars, around any of the Beacons, from several angles.

Thus, the crew of any ship has the option to call up a specific Beacon on the Map, then to select "Background Stars". The first photo of the stars is then brought up to the screen. The crew can then select the specific photo, from their direction. This is how the actual environment of space will appear, to the ship's crew, as they approach the Beacon.

These photos will also be uploaded to the Star-Map Databases, which can then create very realistic displays of the universe.

12b. Specific Protocols for Taking Star Pictures

All photos must be consistent. They must also be accurately labeled. Therefore, we have established a very specific set of protocols for taking the pictures, and labeling them. These details are discussed in other publications, including "Placement of Space Beacons". However, we will provide a few brief concepts below.

12c. Angle Designation of Photos

We need a consistent understanding of the photo angles. Therefore the designation will be as follows.

For the XY direction, we will designate the direction toward the remote planet as the 0° Direction. This is the direction of most interest, as this is the first direction any ship will travel when near these Beacons.

Notice that all photos taken from the ship will be the XY angles. Thus, as the camera makes its circle taking pictures, the direction directly in front of the ship (toward the remote planet) is 0°, while the direction directly behind the ship will be 180°. All other photo angles will correspond to the relative position on that circle.

For the Z angle direction, we will designate the plane of the ship as the 0° Direction. If photos are taken from the Beacon, then the Structural Platform of the Beacon will be the 0° Direction of the Z angle.

12d. Labeling of Photos

In order to properly enter the photos into the Mapping Programs, each photo must be clearly labeled. Remember that the staff entering the information and uploading the photos will be at the Corps. They were not actually there in space when the photos were taken. Therefore, the Measurement Crews must clearly designate the viewing angles of each photo, and everyone must use the same system.

The photo label will begin with the Beacon ID Number. Then followed by the Photo Indication, and Photo Number. Finally, it will state the X and Z angles of the photo. For example:

"A – Hannen – 238; Photo 1; X = 0; Z= 0"

This indicates that we are talking about Beacon #238 of the Hannen Design. The data is a photo. It is the first photo in the series. This photo was taken at angle X=0 and Z=0. This is the camera on the top of the ship facing directly forward, in the direction toward the remote planet.

Note that this information is for the staff at the Corps to enter the information properly, and link all information together. Regarding the actual map user, the perspective will be different. The map user will simply choose menu options. This includes Beacon Number, Beacon Line, or Star Name. The appropriate photos will then appear.

13. Star Locations on the Photos

The Measurement Crew can also identify certain stars on these photos. This identification will be useful to future space travelers.

Thus, the Measurement Crew will highlight these stars, with the star names, on a different set of photos. These will be sent to the Corps Headquarters as a separate collection of images. The staff at the Corps will properly enter this information, in all the databases and maps.

From the perspective of the Map User, he will be able to enter the name of any star, and the proper photo will appear. He can also be already looking at a photo, then tell the computer to highlight the location of a specific star.

The computer produces the highlighted circle around that star on the screen. Data will also appear on the side will show the estimated distance to the star, and other information. This general approach exists for both the Beacon Map and the Star-Map Display.

14. Identification of Other Stars and Celestial Objects

In addition, scientists can also label more stars and celestial objects in the area. Note that this is a task for the scientists, not the Measurement Crew. Also note that this is very low priority.

Thus, while the ship is traveling to the next Beacon Location, the scientists can look over each of the star photos taken while at the previous Beacon. The scientists can take their time, as they name and highlight each celestial object in the photo. They can also estimate distances.

The scientific crew will then enter this information in their Science Database. Note that this is a separate database from the database used by the Measurement Crew.

The scientists can then send this information to the Corps, through the Beacon Line, while stationed alongside one of the beacons. This is of course sending the information collected from the previous beacon...while at the new beacon. This is also lower priority, and only sent after all of the above information has been sent through the Beacon Line.

The scientists can also send the data from the remote planet, after all of the Beacons have been installed.

This information will then be sent along to the Science Division of the Corps, who will enter all the data and all the star photos, into one of the star databases.

Review of Essential Data to Collect
for Each Space Beacon

The Space Beacons will serve as Navigation Guides for all ships passing in the area. Each Space Beacon will emit a specific Beacon ID Signal. This signal is unique to the Space Beacon. Therefore, the navigation crew will be able to receive the specific Beacon ID Signal, and compare to the Beacon Maps. This will help the crew confirm their location, or adjust their course if necessary.

However, to make this work we need two types of realities. The first is the physical reality of each Beacon, placed into position, and emitting a specific ID Signal. The second physical reality, is the set of essential data, which is associated with the Beacon. This data will be entered into the official databases, and then used to create the Beacon Maps.

Some of the initial values will be set into the device itself prior to placement, while others will be adjusted by the Placement Crew. However, the Official Data will always be the data observed and measured by the Measurement Crew, when the Beacon is operational.

In the following sections, we will review each type of data to collect, and the important notes regarding the method or reporting of that data.

1. Beacon ID

The Beacon ID Number is the specific number of the Beacon equipment. This number will be assigned by the Communication Division of the Corps, and will be engraved on an inside panel of the Beacon. All data regarding that Space Beacon will be associated using that number.

An example of the number is: "A – Hannen – 238". In this case, the "A" tells us that it is used as a Main Beacon (in a Beacon Line). "Hannen" is the name of the overall design style of the Beacon. Then # 238 indicates that is unique number 238 of the Hannen Design.

The Measurement Crew will send this Beacon ID Number prior to any other data. This first message of Beacon ID Number is thus telling the Corps that all data being sent in the few next hours will be associated with that Beacon Number.

2. Beacon Line Name

The Beacon Line Name is the name of the Beacon Line between the two planets. This will also become a line drawn and labeled on the Beacon Maps. The Beacon Lines will be named according to specific rules.

For example: The "E2S: Dankhar-Earth Line", for the Beacon Line between planet Dankhar and Earth. And the "S2S Dankhar-Shrivna Line" for the Beacon Line between planet Dankhar and planet Shrivna.

This is the second data fact which the Measurement Crew will message to the Corps Headquarters.

3a-4a Preview: Official Location of Beacon

The Official Location of the Beacon will be designated by the Direct Angle and the Direct Distance. Both of these values will be measured very precisely, using the Path Drawing Program on the Placement Ship.

Note that the Earth as Zero Point will remain the Official Reference Point, regardless of the Home Planet. When we create Beacon Lines from another planet, the first measurements will be from that home planet. However, these will be converted to values from Earth Reference.

Both sets of data will be in all databases and maps. For any database or mapping system, the user can select the Home Planet. All data and maps will be created accordingly.

3b-4b Practical Point: Zero Position of the Earth

The X-Y circle is based on the orbiting plane of the Earth. This is simple enough. However, the Zero Position (for angle and distance) will actually be two different values, resulting in two data sets. The first position is used for practical navigation of ships. The second position is used for scientific comparisons.

The first Zero Value for Earth (angle and distance) is where the Earth is closest, in its orbit, to the Beacon and remote planet. This is the practical angle, as will be used by all ships traveling that direction.

The second Zero Position will be where the Earth is closest to the Sun. This becomes a more absolute geometry, and is used for scientific comparisons. Yet both will be used in the databases

However, the Primary Official Beacon Location will always be based on where the Earth is closest to the remote planet. This is because of the practical use of this reference point. These values provide the angles and distances which are of most practical value to the ship's crew.

3. Direct Angle from Earth to Beacon

The Direct Angle will be given in XY and Z angles. The XY Angle will be determined from a grid we impose on the solar plane. The Z Angle is based on the Earth's equator.

The zero angle of the XY circle will be designated where the Earth is closest to the remote planet. (For the reasons described above). The angle number will increase in the direction of the orbit. Then, the XY Angle to all remote planets and Space Beacons will be determined from this Earth location.

For the Z Angle, this direct angle will be measured from the equator of the Earth, when the orbit position of the Earth is closest to the remote planet. The direct line from this position to the Space Beacon, compared to the Earth's equator, will be the Direct Angle in the Z axis.

These Angles will be known precisely using the Path Drawing Program of the ship. The Measurement Crew will tell the computer to perform the final calculations at the placement location. The Beacon will then be put into proper position, including dish alignments. The Official Beacon Data will then be messaged to the Home Planet.

4. Exact Distance from Earth to Beacon

The second measurement for designating the exact location of the Space Beacon is the Direct Distance. This is the distance from the Earth to the Space Beacon, using the same criteria above.

Thus, when the Earth's orbit position is closest to the Space Beacon, this is the starting distance. It is also the starting location for any ship traveling that route to the remote planet.

When the Placement Crew is at the next Beacon Location, the crew will tell the Path Drawing Program to calculate the Direct Distance (as well as the Direct Angle) between Earth and the Space Beacon. This will be the Official Distance for the Direct Distance, from Earth to Beacon.

Note that this value will be given in kilometers, as this is the most precise unit of distance.

5. Distance Between Any Two Beacons

Knowing the distance between each Space Beacon will be useful for ship navigation. This value will also be determined, exactly, using the Path Drawing Program. Thus, the Measurement Crew will ask the computer for the calculation, and message to the Corps. Note that this distance will also be stated in kilometers.

This data will be compiled, as it becomes available. The measurements will be determined by the Placement Crew. Specifically, the Placement Crew will give the exact distance between the previous Beacon and the Beacon currently being put into position.

This distance, along with many other data, will be sent to the Database Staff at the Home Planet. The staff will then compile the full data of distances, between any two beacons.

For the purposes of routine navigation, the crew will select any two Beacons from their mapping system, which will bring up the distance. The crew will then know where to expect the location of the next beacon.

6. Distance and Angle from Beacon to Brightest Star

The Distance and Angle from the Beacon to the brightest star will have many important benefits. Regarding navigation, the crew can use these values to determine the ship's location.

In addition, this will help future space explorers. Many explorers will deliberately veer from the Beacon Line, to explore new regions of space. Yet they will want to return to the Beacon Line again. Therefore, knowing the directions between the Brightest Stars and each Space Beacon will help the exploring ships to return to the safety of the Beacon Line when needed.

There are other benefits as well. The brightest stars may have other planets to settle, and future Beacon Lines to develop. The brightest stars will also be useful for photon capture, essential for Beacon Operation.

Regarding specific data of the Brightest Star, the Distance and Angle from Beacon to the Brightest star will be measured. These values will be estimated by the Measurement Crew, from the ship, while at the Beacon Location. Note that this can only be an estimate, as we have not yet traveled there. The units will be given as "estimated kilometers" and "astronomical units". The Crew will also highlight and label the specific star being referenced on several photos. This procedure can also be done for several of the Brightest Stars in the region.

7. EM Frequency of Beacon ID Signal

Each Space Beacon will its own unique Beacon ID Signal. This signal is combination of the specific EM Frequency and the Timing of the Bursts.

The specific EM Frequency used will be determined as collaboration between the Communications Division and the company which manufactures the Beacon. This will be engineered into the design, with the ability for space crews to make slight adjustments with a keypad.

The Placement Crew will determine the actual frequency. They will enter the frequency into the keypad. They will observe the exact frequency emitted as the Beacon emits the signal. This observed frequency will be the Official EM Frequency of the Space Beacon.

8. On/Off Timing of Beacon Signal Bursts

The second factor in the Beacon ID Signal is the Timing of the Bursts. This means two values must be set into the Beacon, and recorded as official data. These two items are: the amount of time of a continuous burst ("on"), and the amount of time without any burst ("off").

Each Beacon should be designed with a range of Timing Options. The range is determined by the Corps, yet can be adjusted depending on future decisions. A keypad will allow the workers to choose specific timing for "on" and specific timing for "off".

However, the crew will have been a list of data for the ideal options for this specific beacon. This data list will be compiled by the Communications Division, to ensure that the Timing Options for each Beacon are unique.

Thus, the Placement Crew will set the On and Off timers, then start the Beacon and let the Beacon Signal operate for a few hours. The Measurement Crew will then observe the actual on/off timing of the bursts. This observation, when Beacon is fully operational, will be the Official Timing of the On/Off bursts.

Ideally, the values should be consistent: the suggested timings from the Corps of on and off, the values entered into the keypad, and the observed timing, should all be the same. However, engineering is often imperfect. Therefore, it is only the Observed timing of burst and no burst that will be reported as official values.

9. Initial Intensity

The Initial Intensity is the intensity which is emitted by the Beacon as the Beacon ID Signal.

The specific value of the Initial Intensity is important, because the ship can compare the known Initial Intensity of signal emitted, to the Received Intensity. The difference in intensity can then be used in a simple calculation to determine the distance between the ship and the Beacon.

The engineering of the Beacon Intensity should be built as a range of options. The standard intensity value will be determined by the Communications Division of the Corps, and engraved on a panel door.

The specific Initial Intensity will be set by the Crew, using a keypad. The crew will set the Intensity at the Standard Level, then let the Beacon operate for several hours. The Measurement Crew will then receive the signal, and determine the actual Initial Intensity. This value should be the same as that entered into the keypad, however engineering can be imperfect. Therefore, the observed intensity will be the official value.

In the future, the Corps can decide to change the Initial Intensity of any Beacon. This will be done very easily using the keypad at the Beacon.

10. Background Star Photos

The last data to collect at the Beacon is the set of photos of the Background Stars. The Measurement Crew will take several photos of the Background Stars from many directions. These will be labeled and transmitted to the Corps.

These pictures will be taken according to a very specific protocol. (See the publication on "Placement of Space Beacons)". The collection of photos will then show the viewing of all stars, at all angles, from the Beacon. Additional pictures should be sent of the Brightest Stars. The crew will duplicate some of the photos, then label these stars.

The entire collection of star photos will be carefully labeled and sent to the Home Planet. These photos will then be uploaded into the database systems. This includes the Star-Map Room and Beacon Map.

For the Star-Map Room, the user will select the specific Beacon, and at least one reference star. The Dome Display will then call up the photos taken from that perspective.

For the Bacon Map, the user will search for the specific Beacon, then select "Background Stars". The actual photo will come up, of Beacon and stars, as taken from the Placement Ship. The view can be rotated, which will then bring up different photos, taken from different angles.

11. Distance Between Any Two Planets

For most travelers in space, the final objective will be a Destination Planet. Therefore the crew will want to know the exact distance (and trajectory angle) between the Home Planet and the Destination Planet.

This distance is easily obtained using the Path Drawing Program. Thus, when the Beacon Placement Crew arrives at the Destination Planet, they will again perform the final calculations on the Path Drawing Program. This will give the exact values for Direct Distance, and Direct Angle, from Home Planet to Destination Planet.

Notice that this data will not be collected until the Placement Ship reaches the Destination Planet. Therefore this is the only data which must wait until all beacons have been placed, the ship arrives at the planet.

11b. Two Sets of Distances Between Planets

Note that we should send two distance values, to correspond to different locations of the planet in its orbit.

Both sets of data will have Direct Angle and Direct Distance. However, the first will be based on the Placement Ship intersecting the Destination Planet. The second will be accurate modifications, for where the Destination Planet is closest (in its orbit) to the Home Planet.

The first distance is based on the planet's location when the Placement Crew arrives. Thus, regardless of where the Destination Planet is regarding its orbit, the Placement Ship will deviate from its straight course, to meet the planet at that time.

Also remember that the Path Drawing Program has been in continuous operation since leaving the Home Planet. Therefore, the Placement Crew will use the Program to calculate the Direct Angle and Direct Distance, based on this interception. This value is then sent to the Home Planet, and put into all appropriate databases.

For most purposes, this Distance and Angle is "good enough". However, we should eventually be more accurate. For consistent geometry, we must use the same concept as when leaving the earth. That is: where the Destination Planet is closest to the Home Planet, according the orbit of each. This will be the location for Closest Distance.

Thus, the settlers will perform appropriate calculations, to adjust the first values of Direct Distance (and Direct Angle). This adjusted value will then be sent through the Beacon Line, to the Home Planet.

Therefore, we will provide two sets of data for traveling between planets. Both sets of data will have Direct Angle and Direct Distance. The first will be based on the Placement Ship intersecting the Destination Planet. The second will be accurate modifications, for where the Destination Planet is closest (in its orbit) to the Home Planet.

Additional Data

The ship's crew can send additional data as desired. Most of this additional data will be Scientific Data, rather than Navigation Data. This means that the information is collected by the Scientific Team, rather than the Measurement Crew.

Furthermore, this data, when arriving at the Corps, will be sent to the Science Division for their databases, rather than to the Navigation Offices for the Mapping Databases.

Also note that this data is secondary; only after the other essential data has been sent, can the additional data be transmitted.

Importance of the Data

Each of these Data Items are very important. Though it may seem that there are many data items to collect, it is important that we do collect this data, for each Beacon.

The Beacon as Navigation Guide will only work if the data regarding Beacon Location and Beacon Signal are highly accurate. Therefore, it is essential that the Measurement Crew be well trained in the process of obtaining each measurement. The Measurement Crew must also take their time, to ensure the accuracy of all data.

Similarly, the photos of the Background Stars are just as important. These photos will be used in all the Mapping Programs. Therefore, the quality of the photos, the specific angles for each shot, and the labeling of each photo, must adhere to strict standards. The Measurement Crew must also be well-trained in these processes.

Therefore, in total, the data and photos described in this chapter are essential to the Beacon Mapping Systems; and thus essential for the effectiveness of the Beacons as Navigational Guides.

Chapter 10:

Beacon Map:
Design and User Perspective

Overview of Beacon Maps

The Beacon Maps will show the exact location of all Space Beacons in the network. The data is taken from the Beacon Locator Database, then put into one of the mapping programs. The user can select the Beacon Line, Beacon Number, Brightest Star, or other factors from the data, which will produce the map of Space Beacons in the selected region.

Beacon Map: User Operations

The "Beacon Map" is the mapping system which can show the location of any Space Beacon in the network. The data comes from the Beacon Database, then is displayed as a series of maps. These maps will show the locations of all Beacon Lines, Space Beacons, and Supply Containers.

The layout of the Beacon Mapping Program will have two windows. The main window is the actual Beacon Map, the second window is the Menu and Data Window. This layout will be the same for all viewings of the Beacon Maps.

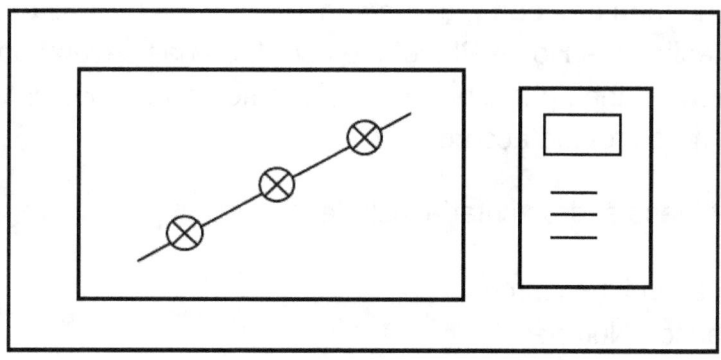

Note that the first page of the Beacon Map will always be the Earth, with each of the Beacon Lines extending from the Earth.

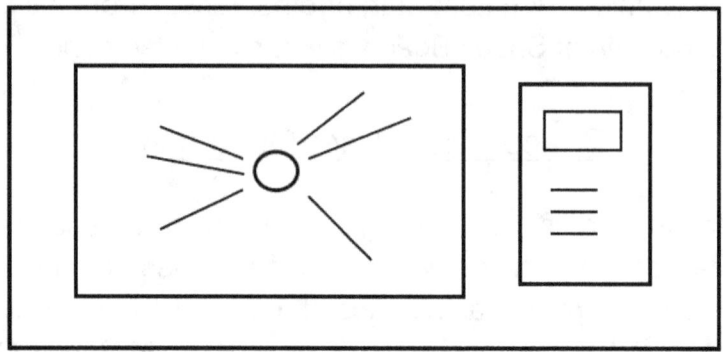

The user can then select the Beacon Line of choice, or a specific Beacon Number. This can be done from the menu on the side.

If the user is on another planet, he can change this view by selecting the "Home Planet" option in the side Menu, and typing the new planet name. He can then proceed with the specific Beacon Line as desired.

After selecting the Home Planet the user then has several options. The Menu Window on the right allows the user to search various criteria. The map will then be created accordingly.

Some of these search criteria include:

- Beacon Line Name
- Beacon Number
- Brightest Star
- Beacon ID Signal Factors

The Beacon Mapping System will then create the Beacon Map. The user can zoom in and out on the screen. Relevant data is listed on the Data Menu on the side.

Maps of Beacon Lines

The Beacon Map is used primarily to show the locations of the Beacon Lines and the locations of each Space Beacon. This will be the primary use of the Beacon Map for most users. In this sense, it will be similar to looking at highway map, and noting the mileage markers.

The user will be able to select the name of the Beacon Line, and the screen will now show only this Beacon Line. The map will thus show the Beacon Line in its entirety, from one planet to the other. Each Space Beacon will be noted with the Beacon Symbol.

Distances will be shown in two axes, on the bottom and the side. Actual space is of course 3 Dimensions. Therefore, the user can change the view to any of two directions. This would be X-Y, X-Z, or Y-Z.

The distances will be given in kilometers, but can also be changed to "astronomical units". The user can move the map from left to right, and zoom in as desired.

Thus, any crew member of a ship can use this system to look at the Beacon Line. They will be able to see the entire Beacon Line, and closer views of any section of line. They will know the distance to each Beacon, from home planet, or from any other Beacon.

Practical Notes on the Beacon Line Maps

Note that the view is in 2-D of 3-D reality. The views can be changed to any other pair of axes, as described above.

However, the ship is likely to not need any other views. The ship will be traveling alongside the Beacon Line, and therefore after one view is selected, this map view is all that is needed.

Also note that the Beacon Line Map shows distances (on each axis), but not angles. The mapping system will automatically orient the map, such that the Beacon Line is drawn left to right, regardless of the actual angle of the beacon line in space.

151

Data Menu of Beacon Line Maps

Data regarding the Beacon Line is presented in the Data Menu on the right. The Data Menu, for each Beacon Line Map, which show all relevant data. This includes: The Names of Each Planet at the end of the Line; Direct Angle from Home Planet to the Remote Planet; and Direct Distance from Home Planet to Remote Planet.

The Data Menu will also show the number of Space Beacons along the Beacon Line, and pull-down menu for selecting a specific Space Beacon. This option will bring up a specific beacon map just for that one Beacon, with appropriate data on the side.

The specific data in the Data Menu will obviously correlate to the information on the screen. Most often, the data will refer to the specific Beacon Line. However, the user can also select a specific Beacon Number; the Data Menu will show all relevant data for that Beacon.

Maps of Specific Beacons

The other main use of the Beacon Map is to locate specific beacons. As the ship passes the Beacon, the crew will desire to see their exact location on the Beacon Map.

Let us review the basic process: When the ship passes a Space Beacon, the ship will receive the Beacon Signal, and the computer will analyze the specific Identification. The computer will then match the Identification Signal to the values in the database. This will result in the computer showing the Beacon Number. This is the specific beacon which the ship is passing at the moment.

With this information, the navigation crew can use the Beacon Mapping System, to know the ship's exact location in space. The user will simply type in the number of the Beacon in the "Beacon Number" search bar. The crew can also go to the Beacon Line (as described above), then select the Beacon Number from the pull-down menu.

The user will also be able to find the specific Beacon by typing in the Beacon Signal ID Factors. The mapping system will then search the database, and pull up the appropriate Beacon.

From this point, a new map will appear, which focuses just on this Beacon. The user can zoom out and in as desired. The Data Menu on the side will then show everything regarding the data of this beacon.

Background Stars

When looking at the Beacon Map for a specific beacon, the user also has the option to add Background Stars. This is an actual photo of the stars, taken by the Placement Ship when in that area.

In most of the Beacon Maps, the "Map" is computer graphics. The Map is a series of Beacon Lines and Beacon Symbols. Each Beacon Line is given a specific color. The background for the basic Beacon Map is a simple black or white.

However when in the Beacon Number Map, we have the additional option of adding the background stars. This is a real photo. Thus, the photo of the actual Space Beacon, with the stars in the background, as taken from the Placement Ship, now serves as the background. Therefore, this is the actual view of the Space Beacon and background stars the crew should see, when they arrive at this location.

Note that the distances will again be shown on each axis. If you wish to see a different view, with stars, we will select this option from the menu. A different photo will be brought forth, taken from a different angle. The computer will then place the distance markers on top.

Other Searches
Brightest Star and Supply Container

The Beacon Map can be used for other needs as well. This includes looking for the Brightest Star and the nearest Supply Container.

Searching for Brightest Star

Searching for the Brightest Star is important when the ship has traveled far from the Beacon Line, and the crew need to find their way back again to the Beacon Line. This is most common for explorer ships, which deliberately wander from the known Beacon Lines from time to time during their exploration.

In this case, the crew will use the Brightest Star as a common reference point. The crew will see the Brightest Star in their viewing area. They will then match this star to Beacons which are also near this Brightest Star. Then, by comparing angles, the navigation crew will find their way back to the Beacon Line.

More specifically, the process will be as follows: The crew will go to the Search Field for "Brightest Star", then type in the name of the star. This can be done in either the Beacon Database or the Beacon Mapping System. The result will be a list of specific Beacons which use that star as their "Brightest Star" in the region. (Remember that these were determined by the crew which placed the Beacon). The user will then select one of the Beacons from this list, for detailed information.

Database of Brightest Stars

If using the database, the data page will show everything about that Space Beacon. This includes the Direct Angle between Beacon and Brightest Star. The navigation crew can then use this information, compared to their current angle to the star, and make the appropriate course adjustments which will lead to the Beacon Line.

Mapping System of Brightest Stars

Using the Mapping System is similar. The user types in the name of the Brightest Star, and the Menu Window offers a list of all Beacons which use this star as their Brightest Star. This list can be further narrowed by choosing the Beacon Line.

The user will then select each Beacon in the results, which will produce the Beacon Map for that Beacon. All relevant data for the Beacon is included on the Data Window. This will include the Angle and Estimated Distance from this Beacon to the Brightest Star.

Thus, using all of this information, the crew can decide which Beacon to head towards, and then the best course to take to arrive at this Beacon. The Explorer Ship will soon return to a known Beacon Line, and can travel this known route quite easily to a destination planet.

Searching for Supply Containers

The crew can do a similar search for Supply Containers. The user selects a location in space, such as Beacon Line, Beacon Number, or Brightest Star. The user then tells the computer to find the nearest Supply Containers. The computer will locate the nearest Supply Containers, then place on the map. The center of the map will of course be the reference point entered for the search.

Thus, the crew can quickly see on the where any Supply Containers are located. The ship will know where to travel for emergency supplies.

Chapter 11:

Beacon Types and General Designs

Overview of Space Beacon Designs

The Space Beacon is designed with multiple purposes. The two main purposes are for use as a Navigation Guide and Long-Distance Communication System. We will now review the basic designs of the Beacons and operations of the Beacon Systems.

Goal of the Beacon Network

The ultimate goal of the Space Beacons is to create a Network of Beacons across the vast distances of space. Using this Beacon Network, ships will be able to navigate easily to any planet. Also, by using this Beacon Network, the people of any planet can easily communicate with the people of any other planet. Therefore, in the following pages we will summarize the main designs of Space Beacons and the Beacon Systems.

Space Beacon as Navigation Guide: Design and Operation

The first purpose of the Space Beacon is to serve as a Navigation Guide. Each Space Beacon will emit a unique signal, which is similar to a lighthouse. When a ship passes the Beacon, the crew will receive the signal, match the signal to the Beacon Map, and therefore know their exact location.

These Space Beacons will be aligned as a series of Beacons between every two planets. Thus, the Space Beacons will serve as Navigation Guides and Mileage Markers as the ships travel between the planets.

Beacon Signal System

The Beacon Signal System operates similarly to a lighthouse. A burst of EM energy is emitted, in four directions, at periodic intervals. Any passing ship within range will receive the signal, and the crew will know they are near a Beacon.

Each Beacon Signal is unique, and is thus associated with a specific Beacon. This allows the crew to match each Beacon with a specific location on the Beacon Map.

The signal is unique by the following factors: EM frequency; amount of time the stream is On; and amount of time the stream is off. These factors are received by the ship, then the computer matches with the Specific Beacon Number. The Beacon can then be displayed on the Beacon Map.

Thus, the navigation crew will go to the Beacon Map, find the appropriate Beacon, then look at the star map. The crew will also use the intensity of the signal to correlate distance from the Beacon.

Therefore, based on the Beacon Signal, the crew knows their exact location in space. They can then maintain or adjust course as needed.

Space Beacon as Communication Relay: Design and Operation

The second important function of the Space Beacon is to serve as a Long-Distance Communications Relay. Using the Network of Space Beacons, we can easily send message between any two planets, across any distance of the galaxy.

The following is a brief description of Beacons as Communication Relay System. Note that a separate publication is devoted to the engineering details of the Communication Relay Network.

Please see the publication: "Communication Systems for Settlements Across the Galaxy" for complete details.

The Beacon Network for Communication

The principle of the Beacon Network is similar to a line of baseball players. One man can only throw a ball so far. Yet if he throws the ball to another man, who throws it to another, and so on, then we can send that ball hundreds of miles.

The same principle exists for our Beacon Network. One message can be sent only so far. Yet we send the message from one Beacon, to another Beacon, to another Beacon, and so on. This allows us to send that one message across millions and millions of miles of space. This is the Beacon Network.

The Beacon Network primarily consists of Beacon Lines. Each Beacon Line is a series of Beacons, which are aligned precisely between two planets. Each Beacon receives the message, boosts the message, then sends the message to the next Beacon. This process is repeated through each Beacon, across space, to the final destination.

The Beacon Relay Systems

The Beacon itself has a pair of Beacon Relay Systems. This is the method for the Space Beacon to receive and re-send the message.

Specifically, the Beacon Relay System will receive the message, boost the intensity of the message, then send the message out again.

The components of the Relay System include: the Receiving Dish, the Intensity Booster, and the Emission Dish.

There must also be a Particle Harnessing System, which provides the supply particles needed to boost the intensity.

The Relay Systems always exist in pairs. This is to allow the communications to travel in both directions simultaneously.

Side Dishes and Ship-to-Planet Communications

Additional side dishes are recommended. These side dishes will communicate with the traveling space ships. When a ship passes the Beacon, the ship will send a message to the side dish of the Beacon. The message will then be relayed through the Beacon Line to the planet.

Messages will be received by the ship in the same way. The planet will send the message through the Beacon Line, then to the side dishes, which any nearby ship can receive.

This system becomes the most effective method for ships to communicate with the planets, in the long-distance travels.

Buffer Systems

When multiple receiving dishes are connected to the same Relay System, then a Buffer System will be installed. This Buffer System will temporarily store one message, as another message is being processed.

Satellites and Surface Antennas

The Beacon Network also includes Geostationary Satellites and Surface Antennas. These become of the connections between planet surface and the Beacon Lines.

Communication Network Develops Over Time

The Beacon Network will grow over time, as we settle new planets and create new Beacon Lines. There will be no end to the development of the Beacon Network. The Network will grow and connect everyone in the universe. Please see the publication: "Communication Systems for Settlements Across the Galaxy" for complete details.

159

Beacon Types: List and Short Descriptions

There are several types of Space Beacons. The primary space beacon is the Main Beacon. All other Space Beacons are variations in design of the Main Space Beacon, and are used for specific purposes.

The types of Space Beacons include the following:

- Main Space Beacon
- Node Beacon
- Solar System Beacon
- Manned Beacon Station
- Harvesting Beacon
- Floating Supply Container

Note that in this publication on Navigation, we will focus mostly on the Main Space Beacon. However all other types of space beacons are fully explained and illustrated in other publications.

Main Space Beacon

The "Main Space Beacon" is the beacon design that will be placed along a Line in space, between two planets. The Main Space Beacon is designed to function as a navigational aid, and to relay messages for long distance communication.

This Main Space Beacon is the primary beacon design which will be used in the network. It is also the basic design, from which all other types of beacons are based.

Node Beacon

The "Node Beacon" is a Space Beacon which is placed at the intersection of several Beacon Lines. At this intersection, the Node Beacon acts as the combination of several Main Beacons.

Furthermore, the Node Beacon is more than just the intersecting beacon. The Node Beacon can direct messages from any one direction, to any other direction. In this respect, the Node Beacon is similar to a switchboard or central railway station.

Solar System Beacon

The "Solar System Beacon" is a Space Beacon which exists within a solar system or at the edge of a solar system. These Solar System Beacons are useful for boosting the intensity of messages, as well as processing messages from multiple locations.

However, the gravitational energy of the central star will affect the position of these beacons, therefore the Solar System Beacons must be designed to self-correct, to maintain their proper position.

It is also useful to have several of these beacons, at different positions relative to the central star.

Manned Beacon Station

The "Manned Beacon Station" is a Space Beacon which is manned by human crew. Any type of beacon, and any location of beacon, can be designed to be a Manned Beacon. However, the ideal choices are Solar System Beacons and Node Beacons.

When referring to the beacons with human crew (versus the regular unmanned beacons), we refer to these as "Beacon Stations" or as "Manned Beacons". Thus, for example, the Node Beacon is automatic, with no human inhabitants, yet the Node Station is a Beacon in the same Node location, yet with full time crew.

These beacons must of course be designed with numerous other features, for the health and general living ability of the crew. The benefit is more advanced management of communications, as well as additional scientific study of the region.

Harvesting Beacon

The "Harvesting Beacon" is designed to capture photons and neutrinos, then send these particles to other beacons. The Harvesting Beacons are located very close to a star, where numerous photons and neutrons are in the area, and can be captured easily.

Using a distribution system, the harvested supply particles can then be given to Space Beacons much further away. This will assist the other beacons in their operations of boosting messages and emitting signal ID.

Floating Supply Container

The "Floating Supply Container" is a storage unit in space. This Container will house essential supplies for space travelers, including food, water, oxygen, and fuel. The Floating Supply Container is not a beacon, yet it will be placed near the beacons, which allows the Containers to be easily found. Thus, in an emergency, the ship can travel to the nearest beacon, and obtain emergency supplies. The Container will of course be replenished by the next ship on the route.

See Additional Publications

The detailed designs for each of these Beacon Types can be found in other publications. These publications include:

- Communication Systems for Settlements Across the Galaxy
- Placement, Maintenance, and Adjustment of Space Beacons
- Particle Harnessing Systems for Space Beacons

Please see these other publications for complete details on the designs of each Space Beacon and their components.

Essential Components in the Space Beacons

There are several Essential Components for any Space Beacon. These components can be grouped into the following categories:

- A. Energy Supply
- B. Photon Supply
- C. Location Guidance System
- D. Message Relay System

These Essential Components are discussed briefly below. We will also discuss the primary devices which comprise each of the systems.

Some of these systems will be illustrated in the next chapter. In addition, each system and all components within each system are fully illustrated in other publications. Please see the many other publications available in this Series.

A. Energy Supply

The Energy Supply is the power source for the entire beacon. The Energy Supply will likely be nuclear. The Energy Supply itself is contained in a modest size box, with electrical wiring extending from the box. The Energy Supply Box will be inserted into one of the Energy Pockets.

There will be several Energy Pockets, each with its own Energy Supply. Each energy supply will also be connected to specific beacon equipment. Together, these Energy Supplies will provide the electrical power needed for all of the components.

B. Photon Supply

The Photon Supply is the supply of particles needed for several beacon operations. This Photon Supply is absolutely necessary.

In order to emit the Beacon Signal, we need photons to emit. Furthermore, we also require photons to boost the intensity of the messages in the relay system. Therefore the Photon Supply is needed for these operations. These photons will come from the nearby space, mostly as background microwave energy photons. If the beacon is close to a star, then these will be photons of a variety of frequencies.

We will also collect neutrinos for the same purpose. The technology is different, yet the goals are the same. We will therefore use a variety of different types of devices to capture and store these particles.

C. Beacon ID Signal / Location Guidance System

The primary use of the Space Beacon is to be a Navigation Aid. The Space Beacon emits a regular burst of EM energy, which is detected by the passing ship. The specific frequency of EM and the timing of the burst, will identify a specific beacon, which can be correlated to a Beacon Map. The ship will then know its exact location in deep space.

C2. Specific Components and Process of Beacon Signal

This publication is devoted to the topic of Navigation Systems for Space Travel. Therefore, we must provide additional information on the technology required for the Beacon Signal System. Thus, the components for the Beacon Signal System include the following:

- Photon Supply
- Frequency Creator
- Intensity Collector
- Burst Timer
- Emitters

Note that there are four emitters, in each direction.

The process begins when the photons from the Photon Supply are taken to Frequency Creator. The Frequency Creator modifies the stored photons to the desired frequency of the Beacon Signal.

The photons are then sent to the Intensity Collector. This device collects the proper number of photons, to have the desired intensity, for the duration of the burst. Note that this value is also multiplied by four, as there will be four emitters.

The Timer is set for specific timings of Burst and No Burst. At the right moment, the group of photons are sent from the Intensity Collector to the four Emitters. Thus, the Beacon Signal is emitted in all four directions, for the scheduled period of time. These bursts are the Beacon ID Signals. These signals can be received from passing ships in any of the four directions, for millions of miles.

Note that this System is referred to by several names, including the "Location Guidance System" and the "Beacon ID Signal".

D. Message Relay System

The second important function of the Space Beacon is to be a Relay System for messages. The "Message Relay System" will therefore receive a message, increase the intensity, then re-emit the stronger message.

The specific components of the Message Relay System include the following:

- Receiving Dish
- Photon Supply Box
- Booster Device
- Emitting Dish

The message is received by the dish, sent through the booster device, then emitted through the opposite dish.

There are at least two distinct relay systems in each Space Beacon. One relay system sends messages from Earth to the remote planet, the other relay system sends messages from the remote planet to Earth.

Note also that in order to boost the intensity of the message, we need a supply of photons. These photons come from space, and are not part of the message photons. Thus, we require the photon capture and storage system. This is described in the next chapter, and other publications.

Optional Systems for the Space Beacons

In addition to the specialized types of Beacons, there are also options for specific devices which can be used on the Beacons. These optional systems include:

- Buffer Systems for Processing Multiple Messages
- Self-Correcting System for Maintaining Position
- Solar Array for Electrical Power or Photon Capture

Buffer System

The Buffer System is used to manage multiple messages being received at the same beacon from different directions. This is recommended in all space beacons.

Self-Correcting System

The Self-Correcting System will maintain the precise location of the beacon in space. Most space beacons will maintain their position naturally. However, some space beacons will need adjustment from time to time. These systems will automatically sense deviations from location, and adjust the position accordingly.

Solar Array

Any Beacon can be installed with a Solar Array. These can be used for harnessing solar energy to power the Beacon. The solar array can also be modified for neutrino capture, which is essential for emitting the Beacon Identification Signal.

Any of These Devices on Any Beacon; See Other Publications

Any of these devices can be installed on any of the types of Beacons. However, some of these systems are suggested for all beacons. Other systems are suggested for certain types of beacons, and when used in certain locations.

Each of these Optional Systems are described in detail in other publications. These include the publications on Communication System, Particle Harnessing, and Placement of Space Beacons.

Chapter 12:

Drawings of Beacon Types and Components

Overview of Beacon Designs

We have developed several Space Beacon Designs which can be used for the Beacon Network. The detailed designs can be found throughout a variety of publications in this Series.

The following pages show the basic designs for the Main Space Beacon. The primary focus areas of the illustrations are for use as Navigational Guide and as Communication System. Additional details can be found in other publications.

Note also that all other Space Beacon Types are based on this design. Again, see the other Publications in this Series for the detailed designs of each Beacon Type.

Structural Sections of Main Space Beacon

There are four structural sections:

A = Platform Structure, with Energy Supply and Photon Capture

B = Location Guidance System

C = Communications Relay System

D = Optional Solar Array

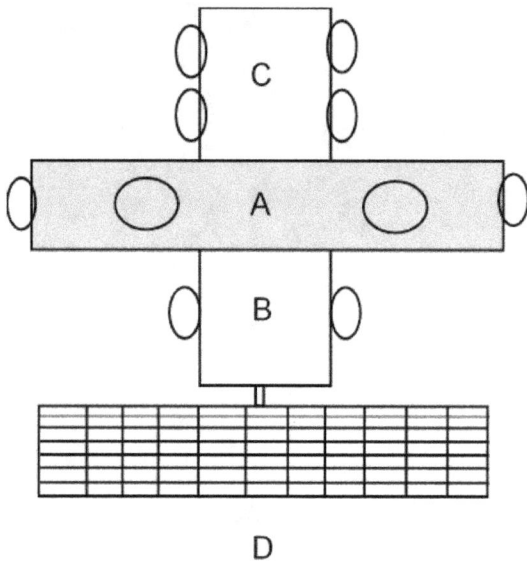

Notes regarding the Dishes

Note that section B (the Beacon ID Signal) has dishes on all four sides, each of which are emitters. Note also that section C (Relay Systems) has two relays. The first is for the messages from the remote planet to Earth, the second is from Earth to the remote planet. Section A has dishes for capturing photons and neutrinos. There are multiple such dishes, on all sides of the Platform Structure.

Location Guidance System:
Schematic Operations

The Location Guidance System is the first function of the Space Beacon. The emitted signal will assist space craft in knowing their exact location. This is useful in deep space, where there are few other celestial objects as guides.

The Space Beacon is powered by nuclear energy, which is stored in the platform. An internal electrical system creates the EM of desired frequency, which is then sent through each of the four emitting dishes.

Each Space Beacon will be identified by a combination of the specific EM frequency emitted, and the timing of the on/off of the emission bursts.

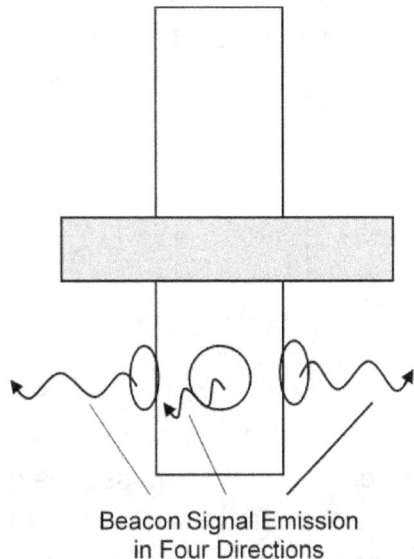

Beacon Signal Emission
in Four Directions

Note the Beacon System will require capturing of neutrinos and photons from the space environment. These neutrinos are used as the photons which will be emitted. For radio wave systems on earth, the neutrinos are carried within the electrical wiring, and therefore are plentiful in each of the emission systems on Earth.

However, in deep space no such wiring exists. Therefore we must capture these neutrinos and photons from space, and convert appropriately to desired photon systems. These designs are demonstrated in the publication on Particle Harvesting.

Communications Relay System:
Schematic Operations

The basic process of the Communications Relay is as follows: The remote settlement sends a message in space to the Relay. The Relay System captures the EM signal, then boosts the intensity of that signal. The message is then sent out again, at a greater intensity, toward the next space beacon. You can see a schematic example below.

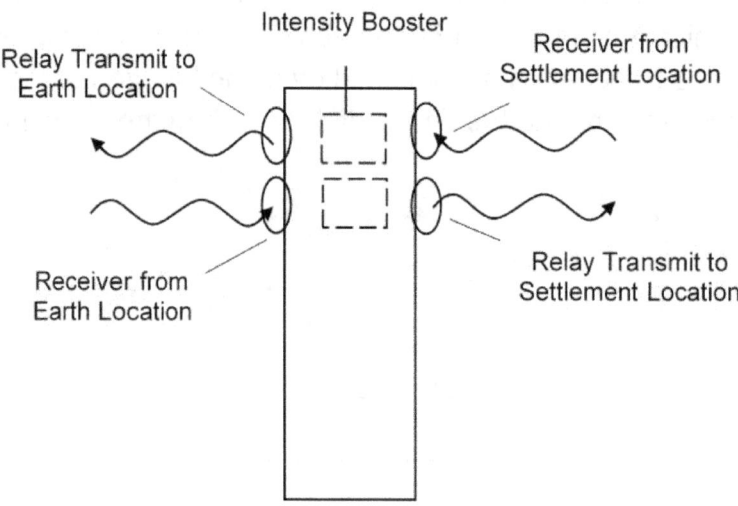

Two Relay Systems on Two Levels: Each Level as To or From Earth

You will notice that there are two separate relay systems, on two different levels. This is to ensure that all messages get to the system.

The first relay system receives messages from the remote settlement, and sends those messages toward Earth. The second system receives messages from Earth, and sends the messages to the remote settlement. Each system operates independently.

Each system will also use a slightly different frequency. This will minimize interference of messages, on each side of the beacon. Thus, the incoming and outgoing, on each side, will not interfere with each other.

Note that the two relay system is important, so that no part of the message gets cut off. Because these messages are for the success of humans living in deep space, it is important that messages are transmitted from beacon to beacon, without interruption. We can ensure this by having separate dedicated relay systems.

Relay Beacons in Series

We will have many of these relay beacons throughout the deep space. Many of these will be aligned in series. This will allow messages to be communicated between remote settlements and Earth most effectively.

In the schematic below, we see several Space Beacons placed in a series. Each one has its own set of receiving dishes, intensity boosters, and transmission dishes. Therefore, by placing a series of these Beacons as Relays in deep space, the communication between the remote settlements and Earth will be more effective.

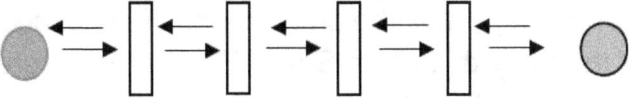

*Note that the schematic is not to scale. It is only for showing the basic process.

Background Science

It is well known that all EM emissions will spread out over distance, and thus the intensity becomes weaker to the detectors which are further away. This is problem when communicating across the vast distances of space. However with a series of communication relays, the message gets a boost in the intensity every few million miles. This will ensure that the messages remain clear over such long distances.

The Beacon Line is useful for sending messages over the very long distances of deep space. Using a series of Space Beacons in a Line, the messages will retain their signal strength. Specifically, message emission sent with a boost in power, as well as being highly focused to the adjacent Beacon. This process will maintain the message signal strength, across the many miles of space. The final signal of messages will therefore be much stronger.

Passing Ships Using the Relay System

In addition to sending the messages from planet to planet using the line of beacons, we will also be sending messages from the passing ships to each planet, using this same system.

This will require adding dishes to the sides of the space beacons. These side dishes will be for the passing ships.

Remember that the primary direction of the relay system is between the two planets. Yet, at the same time, many space ships will be traveling this same route, alongside the Space Beacons. These ships will be using the same Beacons for communicating with the planets. Therefore, the dishes for the passing ships must be on the sides of the beacon.

Thus, each of the two sides of the Space Beacon (90 degrees from the path of planet to planet), will each have two dishes; one receiver on each side and one emitter on each side.

Each of these dishes will be connected to the same relay system, just used less often. Thus, when a passing ship wants to use the Space Beacon for long-distance communication, the crew will pause by one of the beacons, send a message, and the message will be sent to the planet. The planet will then respond, the message will travel across the Beacon Line, and be emitted through the adjacent side dishes.

These ship-to-planet communications will be used quite often. The systems will first be used when the Placement Crew sets the Beacon into position, and sends the Official Data regarding the Beacon.

All future ships will use this system as regular check-in points. It will be standard protocol for all ships to send a status message when passing the next beacon. This allows the Home Planet to know of the ship's location, and that everything regarding the ship's crew is okay.

Note that due to the nature of the emission on the side dishes, many passing ships can listen to the conversations sent through the beacon line. This can be useful, for all space ships to be become generally aware of situations in their region. However, when security is required, the messages can be encoded such that only a specific receiver is able to process the message.

Main Space Beacon:
3-D Drawing of Basic Design

The following diagram is a composite of all the main components of the Main Beacon. The diagram with labels will be found on the next page.

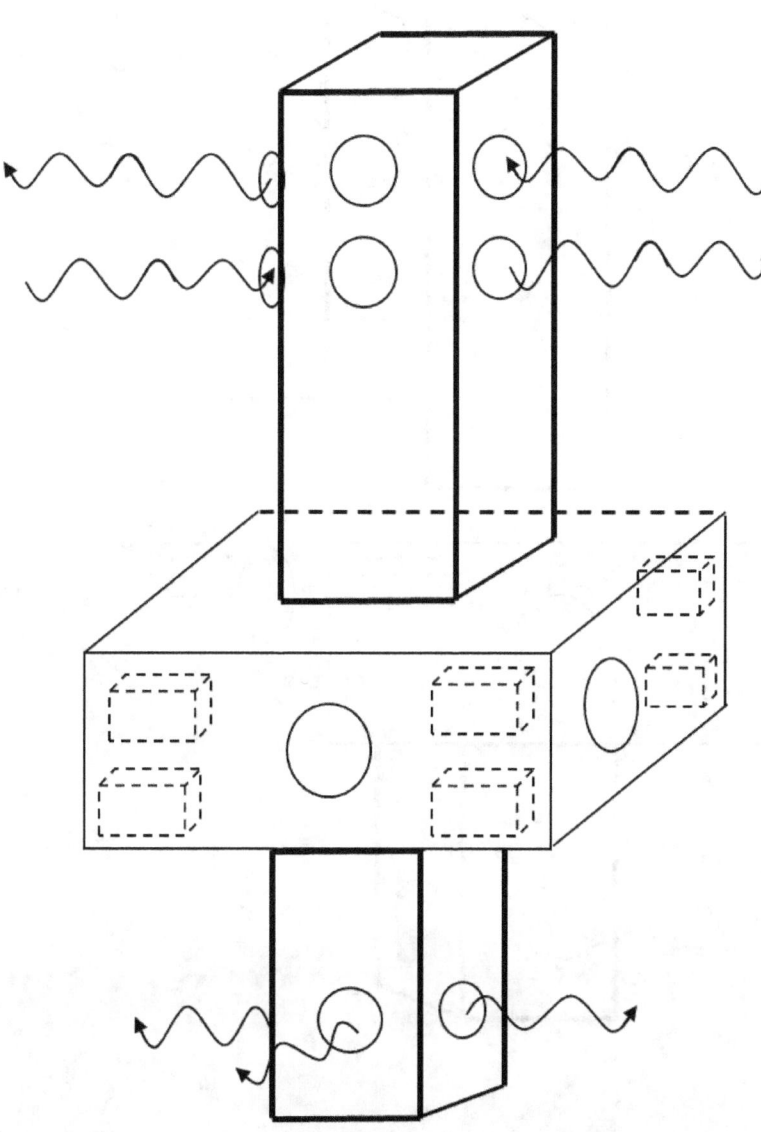

Main Space Beacon:
3-D Drawing of Basic Design: With Labels

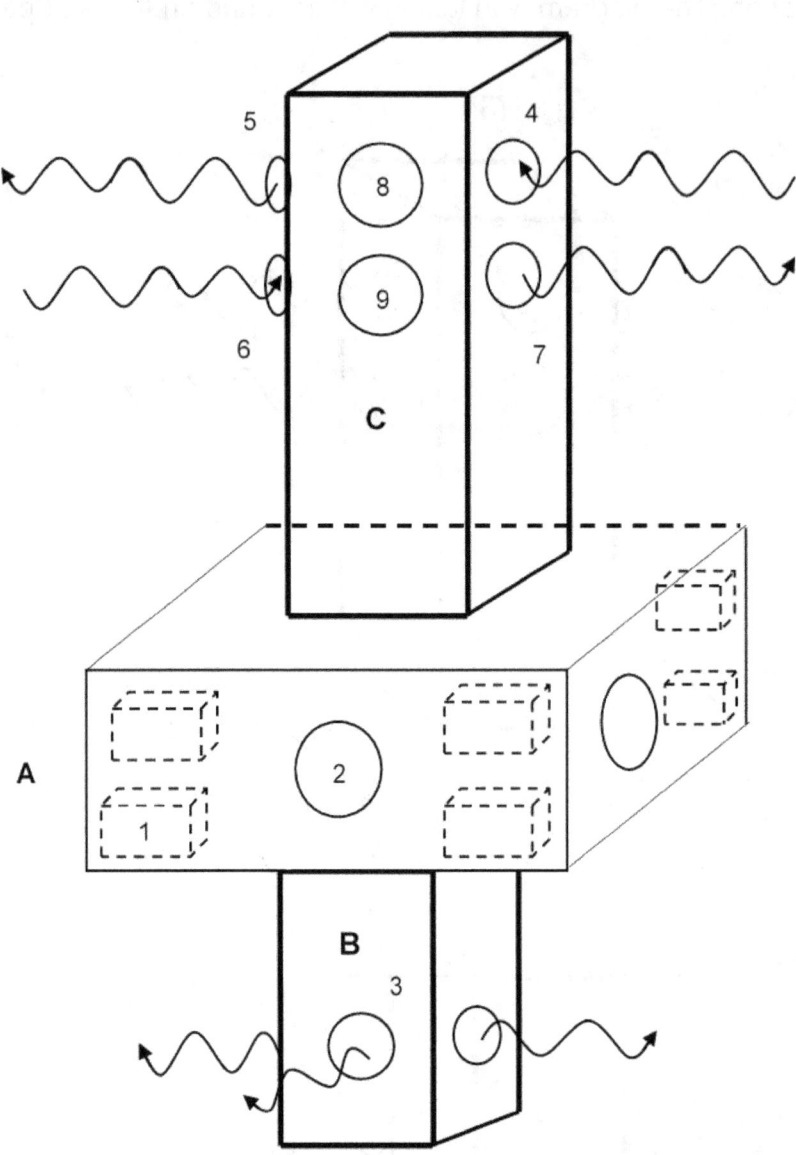

A = Platform Structure

B = Location Guidance / Beacon Signal Area

C = Message Relay Systems Area

1 = Energy Pockets, with Energy Supplies
   (Multiple Energy Pockets in Platform Structure)

2 = Photon Capturing Dish and Storage Box
   (At least one on each side)

3 = Beacon Signal Emitting Dish
   (Emits equally in four direction)

4 = Receiving Dish for Relay System #1

5 = Emitting Dish for Relay System #1

6 = Receiving Dish for Relay System #2

7 = Emitting Dish for Relay System #2

8 = Dish for Receiving Messages from Ships Nearby
   (Similar dish also on opposite side)

9 = Dish for Emitting Messages to Ships Nearby
        (Similar dish also on opposite side)

Node Beacon Drawing:
With Identification Signal Emission

The Node Beacon exists where multiple Beacon Lines intersect. The Node Beacon will manage all the incoming messages without interference. In addition the Node Beacon serves as a switchboard for messages, which can transfer messages to other Beacon Lines.

Yet the Node Beacon should also emit an Identification Signal. This is the same type of signal which serves as the Navigation Guide in all the Space Beacons. The Node Beacon will therefore emit a regular burst of EM, of specific frequency, and specific duration.

The best location for the ID Signal devices in the Node Beacon is the lower half of the beacon (E). The emitting dishes will also be located in that region, on all several sides (F).

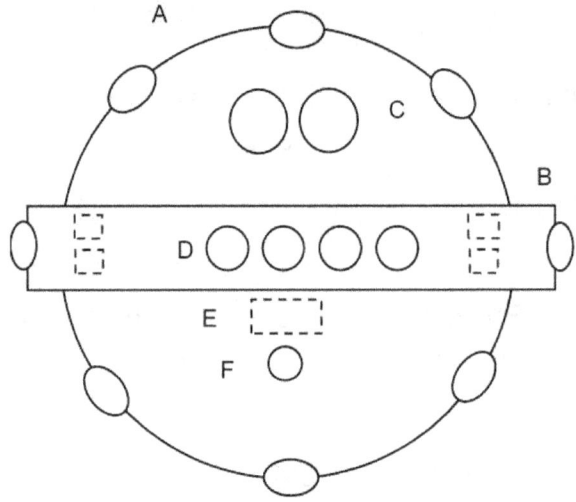

A = Node Beacon Structure

B = Platform Box, External

C = Communication Dishes (Receiving & Emitting)

D = Microwave Photon Capturing Dishes

E = ID Signal Creation System

F = ID Signal Emission Dish

Chapter 13:

Beacon Signal ID and Particle Supply

Overview of Supply Particles

As mentioned above, the Beacon System will require capturing of neutrinos and photons from the space environment. This is necessary for both the Navigation System and the Communication System.

Therefore, in the following pages, we will describe the basic concepts of capturing and converting supply particles. We will discuss the need for these particles and the problem of acquiring these particles. We will then discuss the technologies which have been developed for this process.

Note that these pages will be somewhat brief, compared to the full details of all technologies available. For complete information, on designs and operations related to Supply Particles, please see the publication: "Particle Harnessing Systems for Space Beacons".

Need for Supply Particles in Beacon Signal

The Beacon as Navigation Guide performs its function by emitting photons of a specific frequency. These photons are emitted in regular on/off cycle bursts, and with a specified intensity.

Yet in order to emit these photons, we must have them. We must have photons of desired frequency, to emit when desired. We must have enough of these photons to emit at the regular cycle times, and with the desired initial intensity.

Where do these photons come from? On Earth, we do not consider this question. We are near the sun, and therefore photons are plentiful. However, in deep space, this is a concern. We must carefully capture the photons from space, and harness them for our Beacon Signal.

Need for Supply Particles in Boosting Communications

The situation is similar for the Relay System of the Beacon. The Space Beacon as Communication System performs its function by increasing the intensity of the Message. Specifically, this means adding more photons of the same frequency to the message stream.

Again, where do these additional photons come from? On Earth this is not a problem, as the photons are plentiful. However, in space we must be more strategic in acquiring the photons.

Photon Science: Photon System Structure

In order to properly understand the capturing and conversion systems, we must first understand the structure of Photon Systems.

What we refer to as a photon is more properly called a "Photon System". There are two main components: the particle and the field strings. The central particle is the neutrino. This neutrino provides all of the particle aspects of the photon system. This neutrino acts a high speed train to carry the EM passengers over long distances.

The electromagnetic energy exists as energy strings, which are attached to the neutrino. These are the passengers which ride the neutrino train. Notice also that the amount of these EM energy strings riding on the neutrino is what determines the frequency.

A third component is the gravity strings. These gravity strings hold the EM field strings to the neutrino. These gravity strings also hold each neutrino together; this creates the initial Cluster of photon systems when the photons are emitted.

Photon Science: Creating Photon Systems

Photon systems are created when EM strings attach to the neutrino. The number of EM strings attached determines the amount of "energy" of the EM emission. The amount of EM strings also determines the pulsation frequency and wavelength. (See other publications for complete descriptions and illustrations).

Radio waves, of any frequency, begin with neutrinos inside the electrons. Neutrinos enter the electrons quite often. These neutrinos are everywhere on Earth, and therefore are absorbed easily by the equipment.

Radio emission technology works by adding a certain number of electric and magnetic energy strings to the neutrinos in the electrical wiring. These become the photon systems which are emitted in the antennas. We will perform a similar process in our Space Beacons. We will capture the neutrinos, add the appropriate number of energy strings, and emit the desired frequency.

Note however that there are fewer neutrinos in deep space than on Earth or close to a star, therefore we must develop technologies to attract and capture these particles specifically.

Photon Science: Altering the Photon Systems

Existing photon systems can also be captured and converted to the desired frequency. Remember that the Photon System is essentially a high speed train, with the neutrino particle as the train, and the EM field strings as the passengers. Therefore, we can capture a photon system of any frequency, and convert to the desired frequency. We simply add or remove some of the EM field string passengers, then emit the converted photon system. However, we must first capture existing photon systems.

The Problem of Isolated Beacon in Photon Capture

Let us review the problem. The Space Beacon is far from any star, and is not located on any planet. This means that there are far fewer neutrinos and photon systems near the Beacon.

On Earth, we assume that radio waves can be created anywhere, at any time. We don't consider the need for neutrinos or existing photon systems. They are so numerous, so abundant, that we never have concern. There are two reasons for this: proximity to a star, and the many other objects on earth.

The first reason is that we are close to a star. Therefore, Earth is bombarded with numerous photon systems and neutrinos (as well as other particles) every day. However, the Space Beacon will be much further from any star, and due to the natural spread of particles, fewer will reach the Beacon to be absorbed.

The second reason is that there are numerous objects on the Earth. Each object is absorbing, emitting, and absorbing again. It is similar to a baseball field with a million players and a million balls, where the players constantly throw and catch to each other, in a variety of directions. This is the natural way of photon absorption, and re-emission on the Earth.

However, the Space Beacon does not have this option. Because it is isolated, the Space Beacon does not have other objects to play catch with. The Beacon will become depleted, and unable to emit. Therefore, precisely because the Space Beacon is placed in an isolated area of space, there are far fewer neutrinos and existing photon systems which can be absorbed by the Space Beacon.

The Space Beacon will soon become depleted of these particles, and not be able to emit any signals, unless we develop the appropriate capturing and conversion technologies.

General Overview of Capturing Photons

Having reviewed the basic science of photon systems, we can turn our attention to the process of capturing and converting photons for use as Beacon Identification Emission. The following are some of the main types of capturing systems which will be used in the Space Beacon:

1. Capturing Background Microwave Energies

2. Capturing Photon Systems from Stars

3. Capturing Neutrinos (from Deep Space or Close Stars)

4. Capturing and Carrying Photons using Solar Cells

The Science of
Background Microwave Photon Systems

The most common source of photon systems in deep space is the "background microwave" photon systems. Throughout much of space, there are many photon systems which exist in the microwave range of EM energies. We can capture these microwave photon systems, and convert them to the desired frequency for signal emission.

These microwave photon systems likely came from stars which are very far away, throughout many regions of space. These microwave photon systems continue to travel for many miles, and for many years, after being emitted. Our Space Beacons will take advantage of these photons.

Stream of Photon Clusters, Rather than Sea of Energy

Many people wrongly describe this as a background sea of energy. However, that is not how it exists. The reality is that there are distinct clusters of photon systems traveling through space. These clusters have been traveling in many directions, for many years.

The background microwave energy may appear to be a continuous stream, but in reality it is a series of repeated bursts. This is exactly what we experience with visible light from our sun. We see the visible light as a continuous stream of light, yet in reality it is a rapid series of bursts. The clusters are emitted so often as to appear as a continuous stream of light. This is the reality for visible light from the sun, as well as all the other EM frequencies which are continuously emitted from the sun.

Origin and Presence of Microwave Photon Systems

The original sources of these background microwave photons have not been determined. They likely came from stars which no longer exist.

What likely happened is that there were many stars which emitted a variety of EM frequencies, similar to our own sun. Many of these frequencies were in the EM frequency range. These microwave photon systems traveled billions of miles from the source stars, and over millions of years. The result is that we can today experience these microwave photon systems passing by, at many locations in space.

In addition, it is likely that the other frequencies of EM emitted were quickly absorbed by other objects near to the star, whereas the microwave energies were not. Therefore, over the millions of years, this would leave only the microwave photon systems traveling through space.

Thus, the likely origin of the Background Microwave Energies in space is a set of stars in space, most of which are extremely far away, and some of which no longer exist. The result is many billions of microwave photon systems which continue to travel through space.

We will harness these microwave photons as supply particles for our Space Beacons.

Microwave is Broad Range of Frequencies

Note that the term "microwave" is actually a very broad range of EM energies. Therefore the specific EM energy of the background microwave radiation may be any one of many energies in this range. The capturing technology must be able to appropriately absorb any photon systems within this range.

Capturing the Background Microwave Photons

As described above, there are numerous photons of the microwave energy which are traveling through space. They seem to travel in streams, which indicates a steady source of emission, over a period of many years. The microwave photons also travel in many directions, which indicates multiple sources of emissions, from various regions of space.

Therefore, we have numerous clusters of microwave photons traveling through space. We can use these photon clusters as our primary source of photon replacement, in any Beacon in the galaxy.

For capturing these microwave photons, we will install a set of receiving dishes around the Space Beacon. These dishes are designed to absorb the microwave energy photon systems. Any background microwave photon systems which reach the Beacon will then be captured by one of these dishes.

The microwave photons will then be converted to photon systems of the desired energy. (The energy value then becomes the frequency of photon pulsation). These photons are then emitted as the Beacon Signal.

There best location to place the Photon Capture System is at the Platform Structure.

Platform Structure for Photon Capture

The Platform Structure is the perfect location for the entire capture and conversion system. The Platform Structure is a very large structure (compared to the rest of the Space Beacon), and therefore allows us to place very large receiving dishes on the sides of the Platform. These large receiving dishes will receive large quantities of microwave photons.

These dishes should be as large as practical, considering the size of the Platform walls. This will capture as many photons as possible. Dishes should also be placed on each side of the Platform, which can then capture the photons from all directions.

Therefore, in total, we will be able to use the Platform Structure for the large dishes which capture the microwave photons. The details of the design are shown in other publications.

Notice also that the Platform Structure is large enough to house all of the storage and conversion equipment. Thus, the interior of the Platform will store the photons, convert to desired frequency, before sending to the Beacon Signal Emission.

Design Drawing:
Platform Structure with Photon Capturing Dishes

The drawing below shows the basic design of the Platform Structure, with the Photon Capturing Dishes and the Energy Pockets.

The Capturing Dishes are lined along several sides of the Platform. (These will be on all X and Y direction sides). These dishes will capture the microwave energy photons, and store in the photon storage box. When needed, these photons will be used for Beacon Signal ID and for boosting messages before emission.

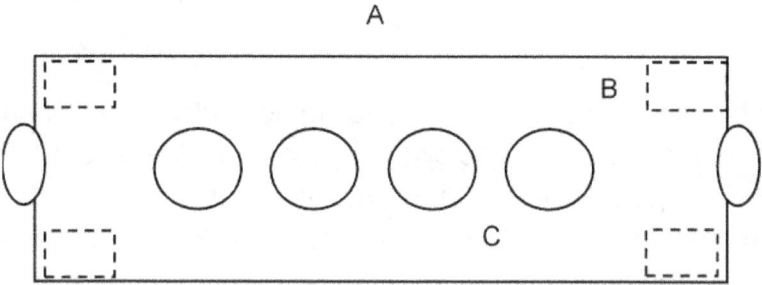

A = Platform Structure

B = Energy Pockets

C = Capturing Dishes, to Capture Microwave Photons

The dashed boxes are the Energy Pockets, which contain the energy supplies. This is usually nuclear energy or batteries. These are best placed on the 8 corners of the Platform Structure. The dishes are then placed between the Energy Pockets, with enough space to easily access Energy Pocket. Note that similar pockets can be made for Photon Cartridge Slots.

Design Drawing:
Photon Capture System in Platform Structure

Every Space Beacon will have a Photon Capture System. This system is necessary to obtain the supply particles needed for Beacon Signal ID, and for boosting messages. These particles come from the microwave energy photons which are constantly passing by the Beacon.

The basic operation is as follows: The photons of space are traveling past the Beacon, from many directions. The Capturing Dishes (H) on the exterior of the Platform Structure are designed to capture the photons of that EM energy.

The photons then travel through each optic cable, to the Channeling Optic Cable (I), as described below. This one cable leads to the Photon Storage Box (J). The Photon Storage Box is essentially a set of mirrors, reflecting the microwave photons inside, and thus keeps the captured photons inside the box.

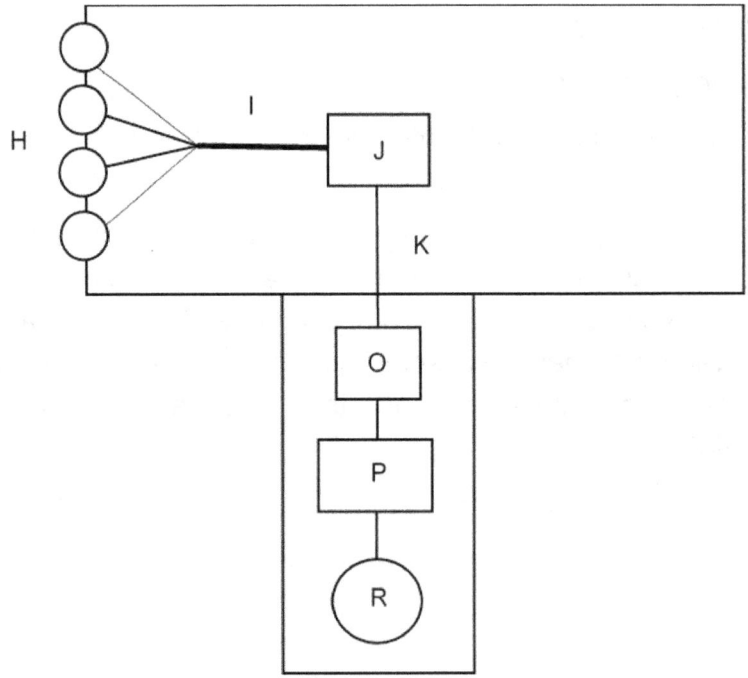

When the photon supply is needed, the exit cable (K) is opened. This is opened for enough time to supply all the photons needed for the next Beacon Signal ID Burst.

This group of photons is then sent to the temporary photon storage (O), until the next time release sends the photons onward. The photons are then sent to the Beacon Signal System (P), which will turn then the supply photons into the proper frequency, then emit the signal in four directions (R) at the proper times.

Design Drawing:
Photon Channeling Cables

We desire many Capturing Dishes on the outside, yet we desire few optic cables entering the Photon Storage Box. Therefore, we compromise by using the Photon Channeling Cables.

In this design, we channel photons from many Capturing Dishes into one cable, This one cable is then connected to the Photon Storage Box. This process is done using the Photon Channeling Cables.

Note that there will be four sets of Photon Channeling Cables, one set for each side of the Platform Box. This will result in four cables entering the Photon Storage Box; yet four is much fewer than the 16 or so cables that would be coming from each Dish to the Box without this merging.

A simplified example is shown below.

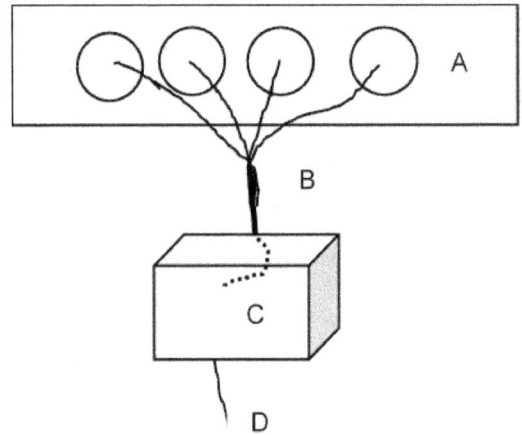

A = Photon Capturing Dishes, Inside View

B = Photon Channeling Optic Cable

C = Photon Storage Box

D = Exit Optic Cable to Beacon ID Signal System

Frequency Conversion Devices:
Microwave to Beacon Signal Frequency

The conversion device will convert the photons of microwave energy into the photons of the desired Beacon Signal energy. This will require adding or removing specific amounts of energy, for each photon system.

Notice that we have an advantage here, because we already know the frequencies of the incoming photons. We know that all the photon systems captured are within a specific range of microwave energy. Therefore, we can be precise in our engineering of the conversion. We will know exactly how much energy to add, or to subtract, to convert those microwave photons to the photons of our Beacon Signal Frequency.

The interior of the Platform will house the conversion devices. These conversion devices will be placed directly behind each receiving dish. The converted photons will then be sent to the photon storage unit. They will remain there, until the Beacon Timer collects the necessary photons for the next Beacon Signal.

Alternately, all microwave photons, from all sides of the Platform, will be channeled to a central conversion device. This will be a larger conversion device, able to process many more photons per minute.

Summary of the Process:
Captured Photons into Beacon Signal

The capture, conversion, and signal emission process will therefore be as follows. We begin with the many photons of microwave energy which are traveling through space. These are photon systems, with the amount of energy per photon that causes the photon to pulsate with the "microwave" frequency.

Many large receiving dishes are placed on all sides of the Structural Platform. These dishes are designed for optimum capture of photon systems with the microwave range of frequencies. Ideally, this range is narrowed as much as possible, knowing the most common range of microwave photons traveling in that area. These dishes will then capture many microwave photons, from all directions.

Behind each dish there will be a Conversion Device. We may also use a central conversion device, which processes the photons from all dishes. The Conversion Devices will therefore convert the captured photons, of microwave frequency, to the photons of Beacon Signal Frequency.

The converted photons will be sent to a Main Photon Storage Box. All converted photons will be kept here. However, some will be taken out, per each time period, for use in the actual Beacon Signal.

This box will be connected to an Emission Storage Box, as the amount of photons for signal emission (see above) will be taken out at one time period, and stored until released to the Beacon Emitter.

The Beacon Signal Emitter will then take those converted photons, from the Emission Storage Box, and channel to all four Beacon Emission Dishes. The Signal Photons will therefore be emitted, as a continuous emission beam, for a specific period of time, with a specific intensity, in all four directions. This is the Beacon Signal which is sent through space.

Chapter 14:

Other Supply Particle Capture Methods

Overview of Other Particle Capture Methods

There are several other methods for capturing the supply particles needed for Beacon Signals. Any of these methods can be used in combination. In the following sections we will provide a brief look at the other options available.

Note that these are brief discussions of each method. The full explanations, and detailed diagrams, can be found in other publications. Please the publication "Particle Harnessing Systems for Space Beacons" for complete details.

The Choices of Particle Harnessing:

We can group the designs of particle harnessing in different ways, based on choices. The first is choice of particle: neutrino or photon. The second choice is the general source of the particles: deep space or from the nearest stars. The third choice is the location of the capturing device itself: on the beacon or near the star.

We can thus summarize the Choices below:

A. Particle Type
- neutrino
- photon

B. Source of the Particle
- deep space
- nearest star

C. Location of Capturing Device
- at beacon
- near the star

Capturing Neutrinos versus Capturing Photons

We can either capture neutrinos or photons as our Supply Particles. However, the technologies will be different.

Let us first review the structure of the Photon System. The particle we refer to as a photon is really a Photon System. It acts as a high speed train, transporting energy from one location to another. The neutrino is the train itself. The Electric and Magnetic Energy Strings are passengers. The specific number of EM Energy Strings determines the total amount of EM energy of the Photon System, and therefore the specific frequency.

Therefore, we can capture either the neutrino or an existing photon system as Supply Particle. We can then change the number of Energy String Passengers as needed to create the desired Photon System.

However, the technologies for capturing and storing the neutrino versus photon will be different.

Capturing Neutrinos

Neutrinos are very small particles, packed with large amounts of energy. The only way to capture these neutrinos is to use electrons. We therefore design an array of electrical wires. The neutrinos hit the array, and are absorbed by the electrons. The electrons then carry their neutrinos to the appropriate destination for storage.

Storing Neutrinos

The best method for storing neutrinos is to keep them inside their electron cage, and in constant motion. In regular circumstances, the neutrino will drive the electron forward for a time, but eventually exit the electron. However, by forcing the electron to move in a complex path, the neutrino remains tossed inside the electron, without the opportunity to escape. The neutrino is therefore stored until needed.

Capturing Photons

Photons are simply neutrinos with EM energy strings attached. These photons are best captured using a set of receiving dishes. In many ways, this is classic radio wave technology.

We choose a range of frequencies to capture. Then we design our receiving dishes to capture and channel those specific frequencies. These photon systems are then channeled through the cables to the storage box, and later to the conversion system.

196

Storing Photons

Photons are best stored using a Photon Storage Box. This box is essentially a box of mirrors, where the reflective material is designed to perfectly reflect the frequencies of the captured photon systems. Thus, the photons are captured and stored until needed.

Source of the Particle

The source of the supply particle can generally be labeled as being from Deep Space, or from the Nearest Star. Of course, all photons and neutrinos originally came from a star. However, for Deep Space, the particles came from stars much further away.

Deep Space Particles

The particles of Deep Space are those which seem to not come from any star. This is because the stars which emitted these particles are so far as to not be visible. Many of these stars no longer exist.

The two main particles of interest in deep space are the neutrinos and the photon systems in the microwave range. Both are plentiful in deep space, though not as plentiful as where close to a star.

Near Star Particles

The particles of the Nearest Star (or Brightest Star) are those which are emitted directly from a known star in the region. This star will emit photon systems in variety of frequencies, and in large quantities. This star will also emit vast quantities of neutrinos.

The advantage is clearly the abundance of supply particles. The disadvantage is the distance from star to beacon. However there are ways to compensate for this, which are explained in other publications.

Location of Capturing Device

The location of the Capturing Device is either at the Beacon or close to the Star. When the Capturing Device is on the Beacon itself, we will either use the Receiving Dish or Modified Solar Array as capturing methods. These will be designed into all Space Beacons.

Where the Capturing Device is near the Star, we use specific equipment to harvest the particles, then send those particles to the beacons. This is known as Particle Harvesting. The technologies are discussed in the publication cited above.

Neutrino Harvesting Design

There are many advantages to using neutrinos. They are plentiful, and can be found everywhere. The neutrinos are also easier to convert to the desired photon frequencies. However, the technologies used to harness the neutrinos as supply particles will be slightly different from those used for harnessing photons. The basic concept is to use the electrons to capture the neutrinos, as illustrated below.

We use a thin sheet material, with a grid of electrical wires. The neutrinos hit the Neutrino Sheet (A), and then enter the electrons. The neutrinos then push the electrons forward. Thus, the electrons with their neutrinos will now travel through the wires.

These electrons with neutrinos will travel through the wires of the grid, to a common exit wire (B). This wire leads a Neutrino Storage Box (C). When needed, the electrons with neutrinos are sent to the Conversion Device (D). The neutrinos are then converted to the photons of desired frequency, and the electrons can be sent back to the Neutrino Sheet.

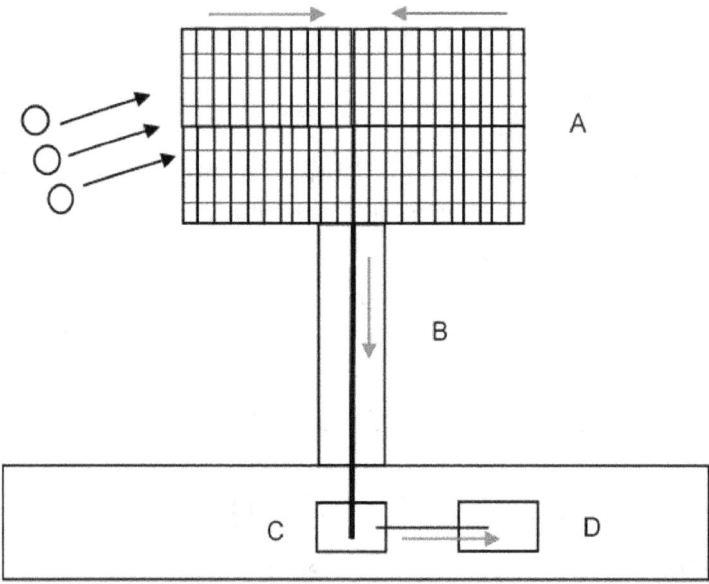

This entire system can be used on any Space Beacon. The system can also be placed on a Harvester near the central star.

Creating Photon Frequency from Neutrinos

When the neutrinos are needed, we simply send a stream of electrons with neutrinos, through the wires, to the conversion devices. This is actually the same process as is done for the creation of all radio wave communications. This is shown in a simplified drawing below.

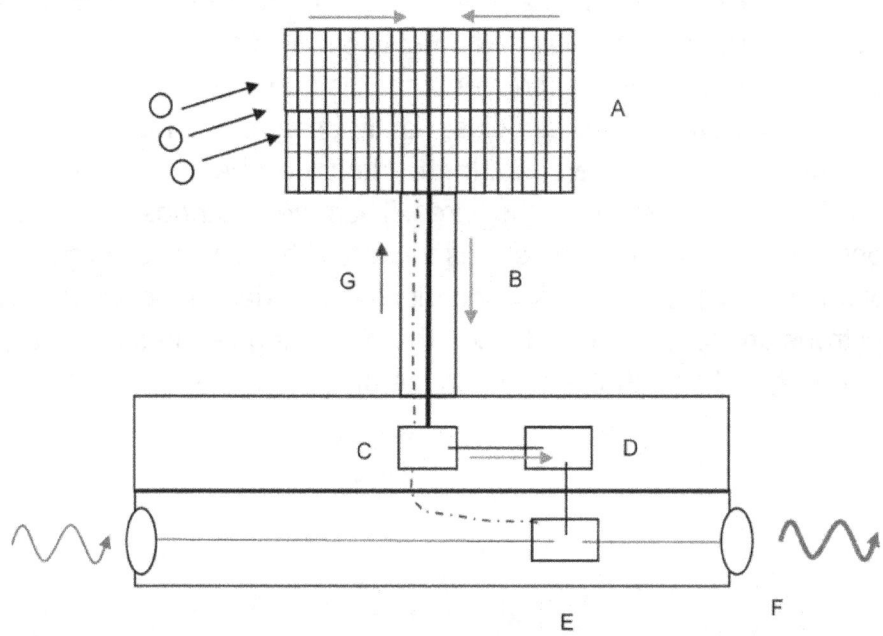

The neutrinos are captured in the electrons of the Neutrino Sheet (A). The neutrinos push the electrons to the common exit wire (B). The electrons with neutrinos are stored in the Neutrino Storage Box (C) until needed. When these particles are needed, we send the electrons with neutrinos through the wire to the Conversion Device (D).

The photons of desired photon frequency are now created. These photons join the original message photons at (E). The message has now been boosted in intensity, and is emitted through the Emission Dish (F).

The electrons are now free of their neutrinos and can be reused. These electrons are sent through a separate wire (G) which takes them back up to the Neutrino Sheet.

Note that there are several components which are not shown. See the publication on Particle Harnessing Systems for details.

Neutrino Sheet versus Solar Array

You will notice that the Sheet Material used for the Neutrino Harvesting looks very similar to the Solar Array. There are similarities. However the technology is different.

They are similar in that we have a thin sheet of material, placed on a post. They are also similar in that both have grid patterns of electronics. However, the Neutrino Sheet is designed to capture neutrinos, whereas the Solar Array is designed to convert photons into electrical current.

The first difference is the particles: neutrinos versus photons. The second difference is the operation. In the Neutrino Sheet, the electrons are already available, sitting in the wires. Thus the neutrinos enter the electrons and then push the electrons forward. This is very different from the Solar Array, which uses a semiconductor material. In the solar array the electrons are only accessed when the photons give the electrons enough energy. Thus, the design and the processes are different.

Combination of Particle Harnessing Systems
and Beacon Operation Systems

There are many ways to combine the Particle Harnessing Systems into one Beacon. Furthermore, we can arrange the layout such that these Particle Harnessing Systems can co-exist easily with the main Beacon Operation Systems. One such example is shown below.

In this layout, we create a much larger Platform Structure. We then build separate posts on the Platform Structure: each Post is the structure for a specific System.

Thus, the Communication Relay System, the Particle Capturing Systems, and the Traditional Solar Array, can each be built on separate posts, essentially parallel to each other.

Increased Size of the Structural Platform

Note that the Platform Structure will be much larger. The length and width of the Platform must be large enough to accommodate all the Posts. The height of the Platform should also be increased in thickness, to provide sufficient support for all components.

Thus, everything is scaled up in size. The entire Space Beacon is much larger, and because of this we can harness more supply particles and more electrical power.

Important to Never Block the Beacon Dishes

Also note that is very important that we do not interfere with the Communication Dishes. The Arrays should never block the signals coming to or from the dishes. This can be accomplished by building the posts to different heights, with the Arrays taller than the Communication Dishes.

Drawing and Viewpoint

An example of this design is shown below. The drawing below shows an example layout of the parallel post designs. Note that in this view we are looking along the line of communication. That is, this is the view if we faced the communication dishes directly. This is important, to show how the other systems can be placed without interfering with the communication relays.

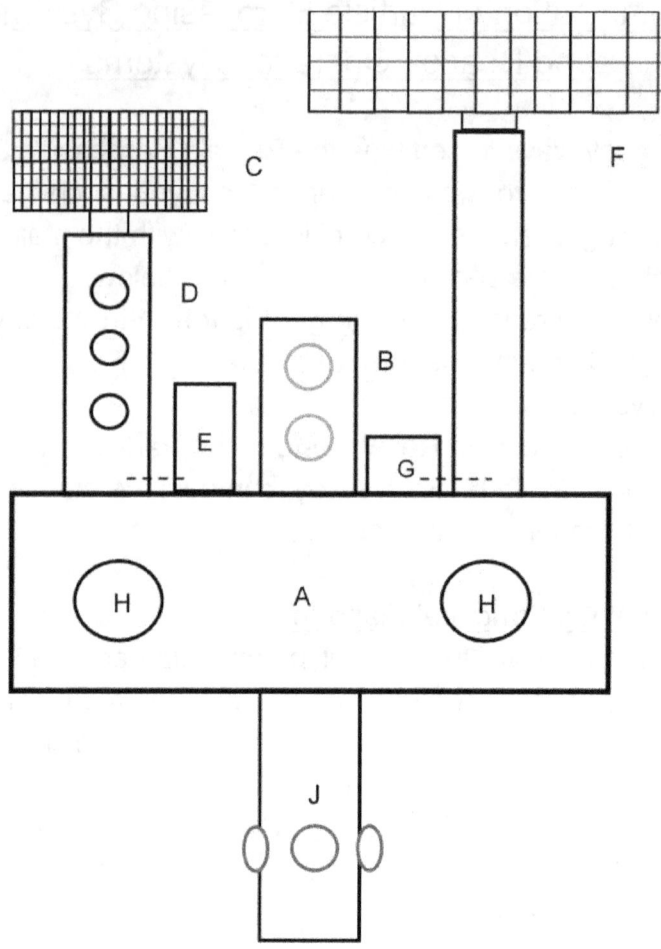

A = Platform Structure

B = Communications Relay System

C = Neutrino Capture Sheet

D = Photon Capture Dishes, on Neutrino Capture Post

E = Particle Supply Container (sturdy container with boxes)

F = Traditional Solar Array

G = Batteries

H = Photon Capture Dishes, Extra Large, on Platform

Top Down View: The Large Square

The Platform is of course a large square. Therefore the positioning of some items can be staggered, or some in front of others, depending on your perspective. The drawing below is a Top Down View, which gives an example of this layout.

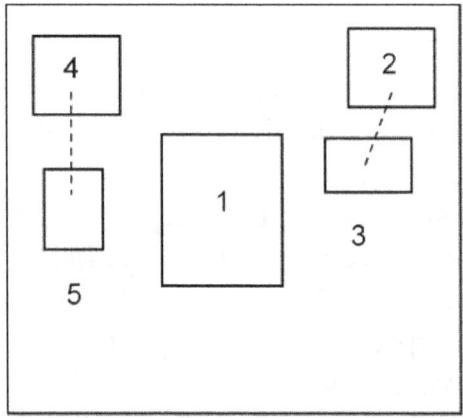

1 = Communications Relay Post

2 = Neutrino Capture Post

3 = Particle Supply Container, with Storage Boxes

4 = Traditional Solar Array

5 = Battery Storage Container

Platform Structure Size

The Platform Structure for this Space Beacon is larger than normal. This includes Length, Width, and Height. There is enough area on the surface to place each of the Posts.

The Platform is also of greater height, which means we can use larger Photon Capture Dishes on the Platform. This will allow us to capture more photons as supply particles at this location. The larger size will also allow us to build more Cartridge Slots (for captured photons), and build more Energy Pockets (for nuclear energy and batteries).

Communications Relay Post

Notice that the Communications Relay Post is designed and built first. This Post is the priority of all posts on the platform. Then the other Posts are designed based on the size of this one post.

Stated another way: the Neutrino Capture Post and the Solar Power Post are both taller than the Relay Post. The Relay Post is designed as needed, the other posts are designed taller to accommodate.

Traditional Solar Array

The Traditional Solar Array in this design is much taller than the other posts. This is important to ensure that the array itself does not interfere with any communication signals.

The array of solar cells can be made as large as is practical. The electrical power generated from the solar array will be sent to a collection of batteries. Thus, the array will recharge the batteries, then the batteries are used for specific purposes. In the design shown here, the batteries will power the communications system.

Also note that the batteries can be sitting on the same platform, protected in a sturdy container. This container of batteries is conveniently located between the solar array and the communications system.

Beacon Signal System

At the opposite end of the Space Beacon is the Beacon Signal System. It is possible to place solar arrays and neutrino capturing systems on that end as well. However, we must be very careful to not block the Beacon Signals. The full intensity of the Beacon Signal is far more important than any particle capture or electrical power we might gain.

Thus, for the Beacon Signal end of the Space Beacon, it is best to use small extensions rather than the parallel platform design. Thus, a small solar array or neutrino capture sheet can be placed on an extension arm, that extends far below the Post of the Beacon Signal System. This will ensure that the Beacon Signals will not be blocked, yet allow some harnessing of particles for our use.

Many Other Layouts Possible

Note that this is just one of the many possible designs. These are shown in the publication cited above. Furthermore, the Particle Harvesting Equipment can be used to deliver particles from any star to the beacons. See the publication on Particle Harnessing for full designs.

Conversion of Particles to Beacon Signal

The Supply Particles will eventually need to be converted into the desired frequency of the Beacon Signal. Specifically this means modifying the number of EM energy strings.

Converting Neutrinos to Desired Signal

This is easiest to do with the neutrinos. The neutrinos begin with no EM energy strings. Therefore we know exactly how much energy add to create the desired frequency. We will set the conversion device, and be certain that the desired frequency is always created.

Note also that this is how most radio signals are created. For both traditional radio signals and our conversion device, the process is as follows: The neutrinos travel inside the electrons to the frequency generator. Energy is added in the desired amount. This energy is attached to the neutrinos as EM passenger strings.

Thus, a photon system, of desired frequency, has been created. The photon system is then launched. This is indeed how our Supply Particles of neutrinos become converted to the Beacon Navigation Signal.

Notice also that the electrons are used again. The electrons which captured the neutrinos are sent back to capture more neutrinos. These electrons never need to be replaced.

Converting Photons to Desired Signal

When using photons as supply particles, the process is similar. However, we must know the frequency of the photons being received. This is why it is best to have a specific conversion device connected to each photon capturing device. The known frequency of capture will be used in the conversion device.

Thus, we begin with a collection of received photons, of known frequency. We then add, or remove, the exact amount of energy from the received photons to create the desired result. This will convert the photons into the desired frequency for Beacon Signal.

Summary of Supply Particle for Navigation

The Beacon as Navigation Guide will perform its function by emitting a regular burst of photon systems, of a specific frequency. However, we must first have photons to emit.

We must capture these photons from space, and convert to the desired frequency. We refer to these captured particles as "Supply Particles". The particles can be neutrinos, or photons of any frequency.

Most of the Supply Particles will come from the neutrinos and microwave energy photons which are constantly flying past the beacon.

Additional supply particles will be captured by pointing the Modified Solar Array and other Dishes at the brightest star.

After these photons have been captured, they will be sent to the conversion device, where these photons will be converted into the desired photon frequencies for the signal emission. The converted photon systems will then be emitted as Signal Identification of the Space Beacon.

Therefore, using a combination of Microwave Energy Capture, Neutrino Capture, and Capture of other Particles from a nearby star, we can harness the supply particles needed, for all the Beacon Operations.

Chapter 15:

Beacon Networks and Navigation

Overview of Beacon Networks

The future of Space Beacons will be an extensive Beacon Network. Throughout most of our discussions on Space Beacons we focus on one Beacon Line. However, as we continue to explore and settle space, there will be many such Beacon Lines, between each pair of planets.

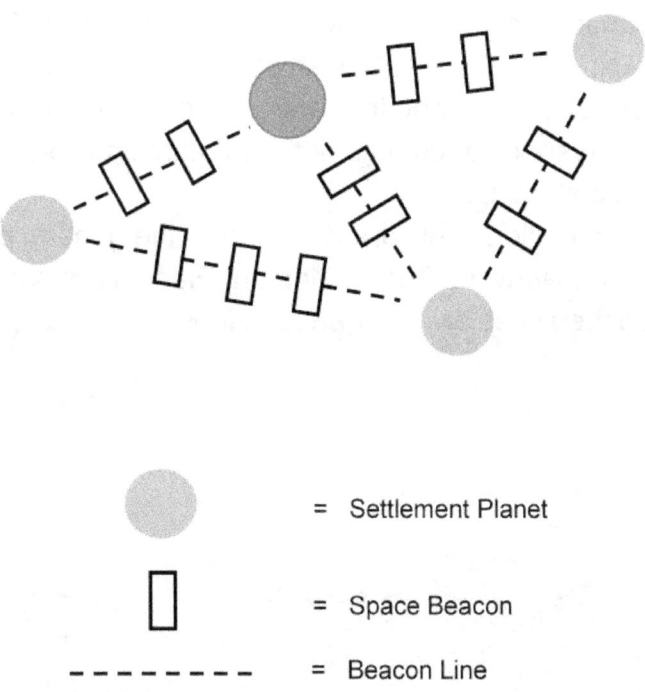

⬤	=	Settlement Planet
▯	=	Space Beacon
– – – – – – –	=	Beacon Line

Note that the drawing is a simple example. The drawing is also not to scale. It is intended only to show the basic layout of the Beacon Network.

Schematic of Advanced Beacon Network

As we evolve our civilization across space, over the centuries, we will also develop numerous Beacon Lines. The space ships will travel alongside these Beacon Lines, and use the specific Beacons as navigational aids. The Beacon Lines will also serve as the method for long-distance communication, from any one planet to another. The ships will also use these Beacons for important communications.

Therefore, these many Beacon Lines will create an extensive Network of Space Beacons. This Network will allow ships to travel across the distances of space without wandering too far from the route. This Network will also allow anyone in space to communicate with anyone else, regardless of the vast distances.

An example is shown below. This drawing shows the general concept of the Advanced Beacon Network. The Settlement Planets are shown as grey spheres. The dashed lines are the Beacon Lines.

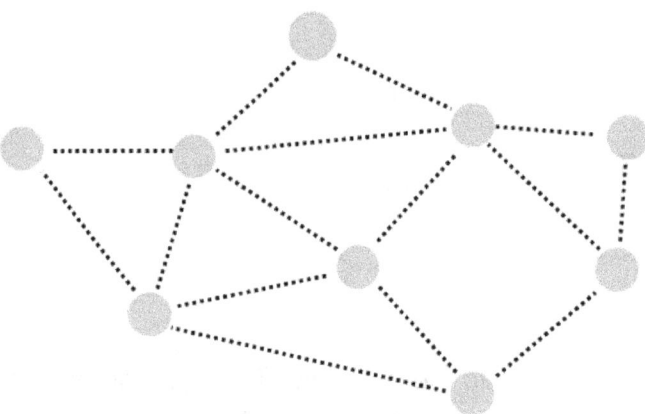

Also note that each planet is in a completely different solar system. The Beacon Lines, though short here in the diagram, actually represent vast distances between planets, with numerous Beacons along each Line.

Ships Traveling Alongside the Beacon Lines

The Beacon Lines will also be the routes which the ships will take between planets. Each Beacon Line is the most Direct Path between two planets. This is why the Beacons, as placed, will be effective systems for communication planets.

This is also why the Beacons will be useful as navigational aids. The ships will travel alongside the Beacon Lines, and therefore pass by each of the Space Beacons. The Beacon Signal, of each Beacon, will therefore serve as a type of mileage marker, and confirmation of their location.

Therefore, each of our space ships will travel along these same routes. Over time, we will develop more space craft, and traveling between planets will become more common. Ships will then be traveling alongside each Beacon Line, at regular intervals.

An example is shown below.

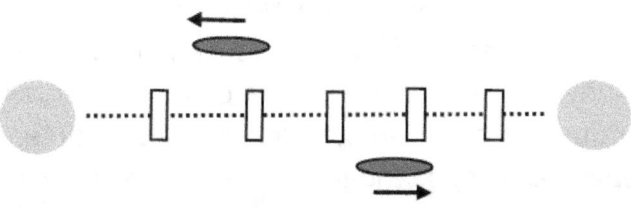

*As with the earlier drawings, the example is meant to show the basic concept, but it is not to scale.

The ideal schedule will be to have at least four ships traveling each route. At any given time, two ships will be at the planets, while the other two ships will be in transit.

This schedule is more improved by having two ships traveling at the same time. This will ensure safety, in case one ship has technical issues, the companion ship can rescue and repair.

Beacon Lines as Navigation Paths

We can therefore see how the Beacon Lines will serve as Navigational Guides. There are several aspects to this.

First, we notice that each Space Beacon serves as a location marker in space. The location remains constant, and is precisely known. Therefore, whenever a ship passes one of these Beacons, the crew knows their location with absolute precision.

Second, the Space Beacons serve as mileage markers. This is similar to mileage markers on a highway. When the ship passes a specific beacon, the crew will know their mileage along the route. They know how far they have traveled from the home planet, and how far they are yet to travel to the destination planet.

Third, the Beacon Line serves as the known path between planets. This is similar to a highway. Thus, as long as the ship manages to travel alongside each beacon, without deviating, then the crew will be sure they are traveling the straight path. Going from beacon to beacon to beacon, the ship will be able to maintain a straight course.

Furthermore, if the ship does not find a beacon where expected, then the crew knows that they have deviated from the course. They will use other computer programs and mapping systems to adjust course.

In this way, the Space Beacons, and the entire Beacon Line, will serve as the perfect Navigational Guide for space travelers. All ships will be able to maintain course, between the two planets.

Beacon Lines as Communication

The Beacon Lines will also be the Communication Systems, for sending messages across the vast distances of space.

This Communication System will be used for both planets and ships. That is, the Beacon Line will relay messages from planet to planet, as well as messages between ships and the planets.

Regarding ships, it is important for the ships to communicate their location and other status reports. Thus, the communication system works together with the navigation system in this way.

Note that the details of the Communication System are explained fully in another publication. Please read the publication: "Communication Systems for Settlements Across the Galaxy" for all the technical details and illustrations. In the sections below, we will highlight aspects of the Beacon Line as communication for the ships.

Operations of Communication System for Ships

Each Space Beacon has communication dishes on four sides. The two main sides face each of the planets. These are used for the communication between planets. The perpendicular dishes ("Side Dishes") face outward toward the passing ships. These are the links between the ship and the planet.

Therefore, when the ship wants to send a message to the planet, the ship will send the message to the facing side dishes. The message is then sent through the Main Dish, and relayed to the one of the planets.

Messages are sent from planet to the ship in a similar way. The message is sent from planet through the Beacon Line, then outward through the side dishes. When the ship is parked next to one of the Beacons, the crew will receive the message.

Protocol for Ship Communications

The Space Beacons will serve as mileage markers and check-in locations. It will be established protocol for each ship to send a status message while passing each beacon.

The message will include: name of the ship, and Beacon Number reporting from. The crew will also report on general status of ship and crew. This is important for the staff at the planet to know of the ship's location, and any other important needs. The ship can then proceed to the next Beacon, where the crew will make the next report.

If the ship does not report when expected, then the planet can send a rescue ship to investigate. This will usually be one of the other ships already in the area.

Options for the Ship to Planet Communications

There are many options regarding ship to planet communication. These options are discussed in detail in the publication cited above.

The first option is which planet to send the message. The suggested design is that the Beacon automatically sends the message to the closest planet. However, switching paths is easily designed.

The ship has the option of waiting for a reply, or continuing onward. If the ship is simply sending a status update, then the ship does not need to wait. However, if response is desired, then the ship can wait.

Ships Listening to Communications with Other Ships

Notice also that the reply message from planets will generally be broadcast through all side dishes, of all beacons...as well as being relayed down the Beacon Line. This design is simpler. It will also provide general messages to all other ships in the area.

It is good for all ships to know of the locations of other ships. It is also important for the crews to be able to assist each other when needed, especially during an emergency.

However, we can easily design systems to send the message to only specific side dishes, and reach only the waiting ship.

These are just a few of the many options available. The main concept is that we are able to communicate effectively between any ship and either planet on the route. If the ship needs assistance or advice, the beacon communication system will make sure this happens efficiently.

*For full details on the Communication System, see the publication "Communication Systems for Settlements Across the Galaxy".

Nodes of Beacon Networks

There will be many intersections of the Beacon Lines. These intersection points are the "Nodes" of the Beacon Network. More precisely stated: a Network Node is any region in space, where multiple Beacon Lines intersect.

Wherever there is a Node, we will place a Node Beacon. A simple example of the Node Beacon, is shown below. Notice how multiple Beacon Lines intersect at the same location.

The Node Beacon is designed to manage all the messages which arrive at this location. The Node Beacon will also be able to transfer messages from one Beacon Line to another.

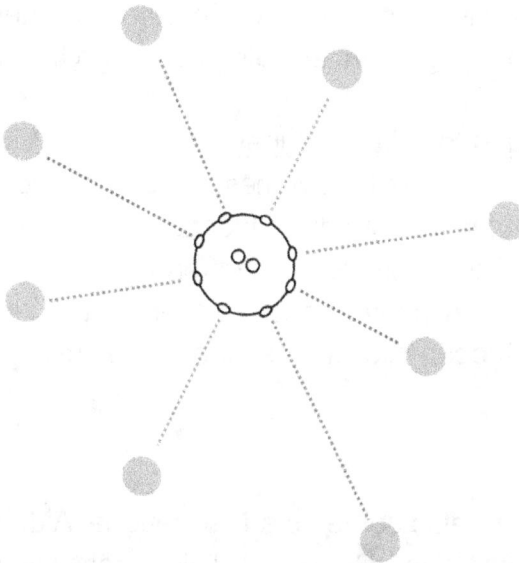

Also note that this is a schematic diagram. It is not to scale. The Node Beacon is emphasized and therefore larger. The planets are further away, and therefore appear smaller.

Node Beacons as Special Beacon Designs

The Node Beacon is a special type of beacon. This is because the Node Beacon will perform many functions. These functions include:

1. Beacon as Communication Relay

As Communication Beacon, the Node Beacon will act as relay for the messages along the Beacon Line. Furthermore, the Node Beacon is a combination of Beacons. That is, the Node Beacon acts as several regular beacons, one for each Beacon Line, now merged into one device.

2. Traffic Control System

In addition, the Node Beacon will act as a traffic control system. Notice that multiple messages will be intersecting. Therefore we need a device which will hold some messages, while letting others go through.

3. Messages Transferring to Different Lines

Further, the Node Beacon will allow messages to change direction. Rather than just continuing on the same direction, along the same Beacon Line, the message can be transferred to a different Beacon Line. This is similar to a central train station, where passengers get on a different train. In this way, the Node Beacon allows messages to be travel in many possible directions.

4. Navigation Aid

The Node Beacon will also serve as a Navigational Aid, where the Beacon emits its specific Beacon ID signal. This process is generally the same as any other Space Beacon.

Therefore, for all these reasons and more, the Node Beacon will be designed differently than the Main Beacons.

*For details on the Node Beacon designs, refer to the publication on Communication Systems cited above.

Navigation and the Node Beacon

The Node Beacon will serve as a Navigational Guide for the ships. Yet because there are multiple Beacon Lines intersecting, the Node Beacon as Navigation Guide will have additional features.

Multiple Routes Intersecting

Remember that a Beacon Line is many things. This includes the line of Beacons as Communication Relays, the Mileage Markers of the Beacon Signals, and the Path of the Direct Route between Planets.

Therefore, when the Beacon Lines intersect, the routes for the ships are also intersection. Ships will traveling across the Node Beacon, at various angles, which correspond to their Beacon Line Path.

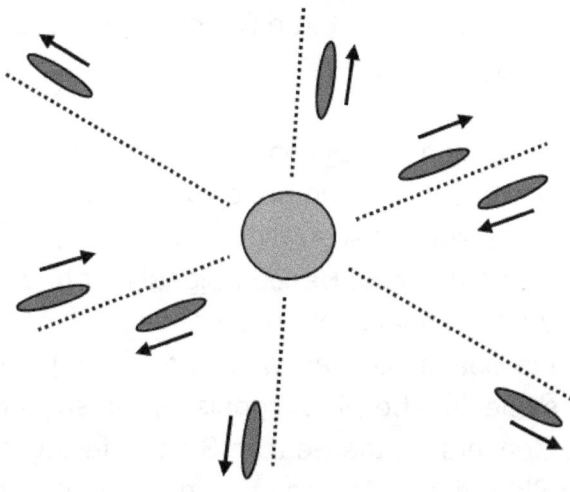

This brings us practical issues to consider. The first is direction of the Beacon Signals. The second is the protocol for ships passing this region at the same time.

Beacon Navigation Signals in Multiple Directions

The Node Beacon will require Beacon Signals in multiple directions. This is necessary to act as navigation guides to ships traveling on each route. Thus, just as we have multiple communication dishes on various sides of the Node Beacon, we must have Beacon Signal Dishes on various sides of the Beacon.

For the Main Beacon, we emit the Beacon Signal in four directions. Two of these directions face the passing ships. These are the main Beacon Signal Directions which the passing ships will receive.

The other two directions will face the planets, and therefore essentially face the oncoming ships. The ship will be able to receive some of this Beacon Signal (as cone shape spread) while nearing the Beacon.

We will do the same with our Node Beacon. For each Beacon Line, we will have essentially four Beacon Signals, emitting in each direction regarding that Beacon Line. This is multiplied by each of the Beacon Lines which intersect. Thus, if there are three Beacon Lines intersecting, we will have twelve Beacon Signal Emitters.

Practical Design for Beacon Signals on Node Beacon

However, we can reduce the number of Beacon Signal Emitters with careful design. The absolute Beacon Signals will always be facing the ships. Yet these "Forward" facing Beacon Signals for one Line can in fact be "Side" facing Beacon Signals, for another Line.

For example: the Beacon Signals which are Facing the ships traveling Line A, will at the same time be Side Signals for the ships passing alongside Line B. Conversely, the Beacon Signals facing the ships on Line B, will also be the Side Signals for the ships passing alongside Line A.

This will eliminate duplication of Beacon Signal Placement, while providing essentially the same result.

Of course the Beacon Lines will rarely intersect at perpendicular. Yet this should not matter, because the Cone Spread will ensure that enough of the signal will reach the passing ship. The passing ship will receive enough of the Beacon Signal for the ship to recognize this Node Beacon.

Furthermore, the computers will be programmed with the geometric data regarding Beacon Signals to Beacon Line, for each of Beacon Signal to Beacon Line option. The computer program for the Beacon Map will automatically take this into account.

Protocol for Ships Crossing the Node Beacon

The Node is a location where multiple Beacon Lines intersect. This also means that multiple Routes for the ships will intersect. The practical result is that ships traveling from different direction will pass this intersection point, often at the same time.

Therefore, in order to avoid complications, we must establish a protocol for the ships which cross this same intersection point.

The simplest method is to establish height levels, relative to the Node. Thus, as the ship approaches the Beacon, they will move up or down, to their designated level. The ships can then pass the intersection point at the same time, without any chance of collision.

Simple Protocol Example for Passing Node

A simple protocol can be established as follows. Ships traveling the route of Beacon Line A will pass above the Node Beacon. Ships traveling the route of Line B will pass below the Node Beacon. Ships traveling the route of Line C will travel alongside the Node Beacon.

This is of course a simple example. We can further establish specific height ranges, for the ships of each Beacon Line. Furthermore, the Node may have many more intersecting Beacon Lines. We can establish different height ranges for each of these.

Thus, it does not matter how many Beacon Lines intersect, or how many ships pass the intersection point at the same time, using this system all ships will travel far from the others. Establishing the Height Ranges for each Beacon Line as Route, we will ensure safety clearance for all.

Manned Node Stations

The Node Beacon can also become a Manned Node Station. The Node Station is a fully manned station, and can therefore perform many additional functions. For example, the staff can perform scientific studies. The Station can serve as rest stop for the passing ships. The Beacon Operations can also be managed more effectively.

However, there are many additional technical factors to consider. The Node Station must support human habitation, and have adequate supplies for long periods of time. The Station is also must larger. These details and many others are fully discussed in the publication: "Communication Systems for Settlements Across the Galaxy."

Solar System Beacons

The Solar System Beacons are similar to the Main Space Beacons of the Beacon Line. However, the Solar System Beacons are placed within the boundaries of a solar system. There are several benefits, yet there must also be modifications.

We can further categorize these Beacons as those 1) at the Edge of a Solar System, versus 2) those beacons Within a Solar System. The designs will be similar, yet the purposes will be different.

The beacons within the solar system will transmit the messages between planets within that solar system. These beacons will also create a type of beacon line within the solar system.

The beacons at the edge of the solar system are those which send messages to deep space. These Edge of Solar System Beacons are useful for giving one final push of intensity, before the message is sent along the Beacon Line of deep space.

The greatest benefits of the Solar System Beacons are for stronger communication signals, and additional scientific studies of the region. Regarding navigation, the design of the Beacon Signals will be essentially the same as any other Beacon.

However, there are additional factors to consider, and modifications which must be made. The primary factor is that the position of the Beacon, being part of the solar system, will be affected by the gravitational pull of the central star and the nearest planets. Therefore we must design around this. Yet there are benefits as well, such as getting greater supply of photons, and ability to boost the messages further into space.

All of these details are fully discussed in other publications.

Satellites for Beacon to Planet Communications

We must also mention the Satellites in the Beacon Network. The satellites will send message from the planet to the nearest beacon. The satellite will also receive messages from the nearest beacon and send down to the planet. There will be corresponding satellites above each settlement planet.

These satellites are standard technology, without any significant developments. We only ask that they be dedicated to the Beacon Lines.

We further recommend using several satellites. This will allow at least one satellite to receive messages from the Beacon Line at any one time.

These details are further discussed in the publications cited above.

Chapter 16

The Universe Grid System
and the Numbered Space Regions

The Universe Grid System

Overview of the Universe Grid System

In addition to having very specific locations for objects in space, we can also divide the universe into regions. This can help us with our practical applications of space exploration.

It is common to divide territories into broad regions. For example, many government agencies place a general grid over their nation, and divide into sections. Each section is numbered. Then individual managers are given jurisdiction of the activities (related to that agency) in that region.

We can do the same with the Universe. We will divide the Universe into a Grid System. Each Grid will be given a set of numbers. This will make our discussions much easier. We can then refer to specific Regions of space, and point to these Regions on the Universe Map. All discussions of space exploration, star systems, beacons, and more will then be organized based on Space Region.

The Universe Grid System and Cubic Space Regions

The entire universe will be divided into a set of Space Regions. Each Space Region will be uniformly square, with dimensions of 3 Light-Years on each side. These Regions will then stack on top of each other (in our mapping system) as a set of many blocks.

The computer program will show a stack of cubic Space Regions, as many blocks stacked together. We will then select the Space Region of interest, the other blocks will disappear, and leave only the Space Region desired on the screen. All of our discussions (with the other scientists in the room) will then focus on that Space Region.

Earth as Center Block

The Space Region which contains our Earth will be the Center Block of the entire Universe Grid System. This is because Earth is our home. It is where we are starting from. We may like to start from the true center of the universe, but this is not practical.

The First Space Region

The first Space Region will be a cube of space around the Earth. Furthermore, the Earth will be at the very center of this cube.

Remember the cube of each Space Region will be 3 Light-Years per side. However, we are placing Earth at the center of the First Space Region. This means that the boundaries of the First Space Region will be 1.5 Light-Years from Earth, in each direction.

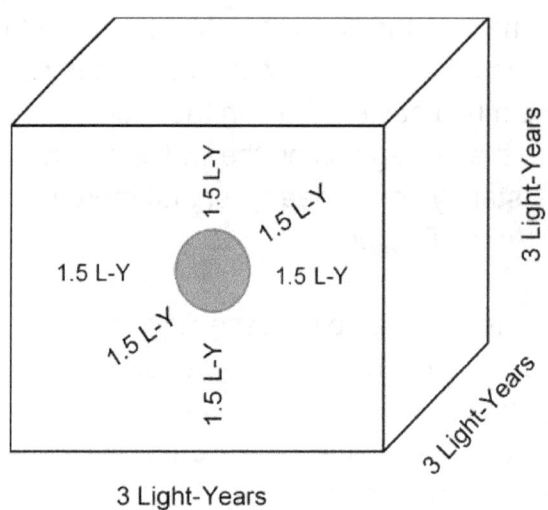

First Space Region
Earth as Center

Numbered Space Regions

Numbering the Space Regions

The Space Regions of the Universe Grid will be numbered according to three numbers. Each number represents the location in the X, Y, and Z directions; with the Earth's Space Region at the center.

For example, a Space Region has number: "4, 2, 5". This means that be Space Region Cube is 4 cubes in the X direction; 2 cubes in the Y direction, and 5 cubes in the Z direction. Therefore, using this numbering system, we can specify each Space Region Cube in the Universe Grid.

Earth's Space Region is 0,0,0

Notice that the Space Region with Earth is given the position of 0,0,0 in the Universe Grid. All other Space Regions are built around this First Space Region.

Universe Grid System, Set #1

The following illustrations show some of the Space Region Cubes, with exact Numbering, within the Universe Grid System.

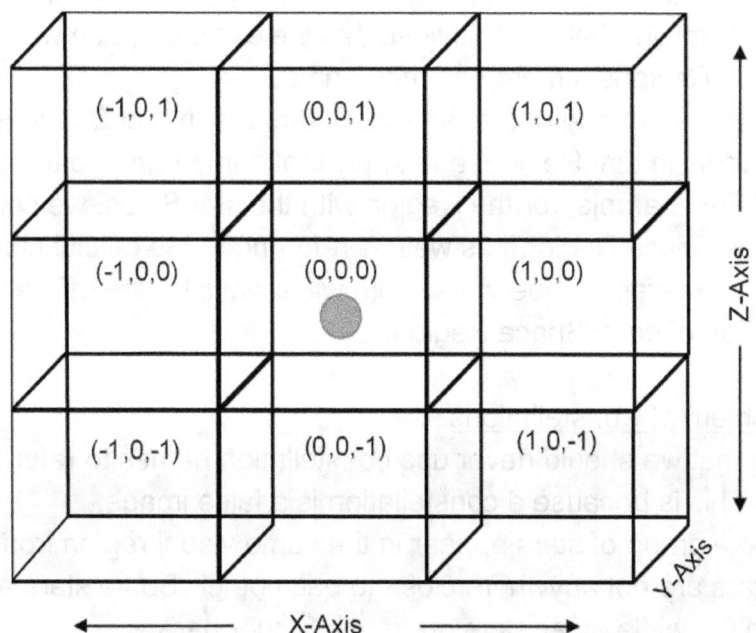

Universe Grid System, Set #2

This diagram focuses on the Space Regions in the Y directions. The Numbers are exact as should be used in the System.

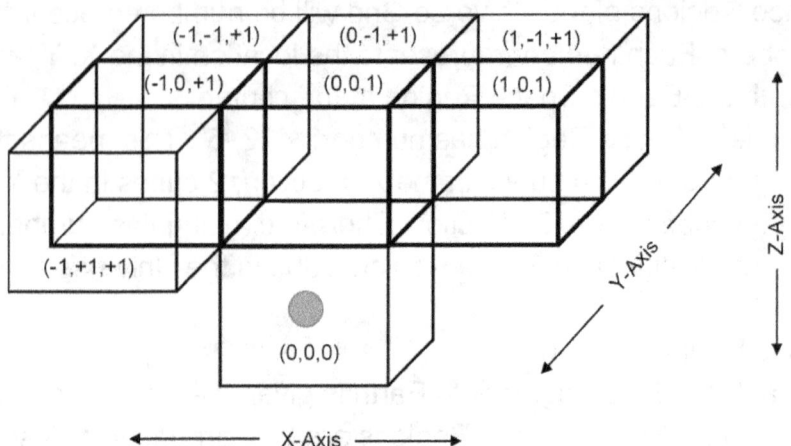

Star Names and Space Regions

Star Names and Nebula as Common Names for Space Region

The Official Designation for each Space Region will be the step numbers explained above. However, as we explore space we can also give these Regions a more common name.

We can select one of the Brightest Stars within that Space Region, or a major nebula in that Region, and apply that same name to the Space Region. For example, for the Region with the star Sirius, we can refer to the "Sirius Space Region", as well as referring to the official numbers. However, the step numbering system will always be the official designation of each Space Region.

The Problem of Constellations

Note that we should never use constellation names to refer to a Space Region. This is because a constellation is a false image.

While a group of stars appear in the same visual region from Earth, these stars are not anywhere close to each other. Some stars are much closer to us, while other stars are much further back.

Thus, many stars of the same "constellation" will actually be in different Space Regions. It is for this reason that we should never use the names of constellations for naming the Space Regions.

Larger Space Regions

Overview

We can extend the concept of Space Regions, by making larger groupings of Space Regions. For example, we can group 1,000 Space Regions together, to create a type of super-region.

Deca-Cube Region (DCR)

The first of these Larger Space Regions will be the DecaCube Region. The DecaCube Region is a set of Single Space Regions, which is 10 cubes on a side. Thus, the DecaCube Region is:

- 10 Cubes x 10 Cubes x 10 Cubes
- 30 Light-years x 30 Light-years x 30 Light-years

Numbering the Deca-Cube Region

The Deca-Cube Regions will be numbered in the same way as the Single Space Cube Regions. However, a prefix will be used to designate that we are referring to the Deca-Cubes of the Universe Grid.

For example: "Deca-Cube Region (2, 7, -3)"

This can also be abbreviated as: "DCR 2,7,-3"

Larger and Larger Regions of Space Cube Groupings

This concept can be continued, as needed, to discuss larger and larger regions of the universe. For example, the next step larger from the Deca-Cube Region will be the Centa-Cube Region.

The Centa-Cube Region (CCR) will be 100 SCR cubes per side:

- 100 Cubes x 100 Cubes x 100 Cubes
- 300 Light-years x 300 Light-years x 300 Light-years

Single Space Cube Regions versus DecaCube Regions

For clarification in our reports, we must specify which Region Size we are referring to. The standard will be the Single Space Cube Region (SCR). The next size upward is the Deca-Cube Space Region (DCR). Other sizes will proceed from there, such as the Centa-Cube Region (CCR), and so on. The specific Space Region will always be designated first with the Prefix (SCR, DCR....) then the Step Number (as X, Y, Z) relative to the Center Region.

Practical Applications of the Space Regions
of the Universe Grid System

Overview of Practical Applications for Universe Grid System

There are many practical applications for using the Universe Grid System. Indeed, the reason for adopting the system is for the many practical applications. This includes applications related to navigation systems, as well as general management of space travel.

Space Regions and Star-Maps

The Star-Maps are based on photos collected from previous ships on their journeys. Similarly, the purpose of the Star-Map Room is as visual guide to the reality of stars during a ship's travels.

The accuracy of the Star-Maps depends on the number of photos in the database. However, at the same time, with more photos to search, the computer will take a longer time to select the photos for the desired display. We can make a compromise, by including only those photos which the ship is likely to need. These photos can be grouped by Space Region. The ship will then load into the database only the photos of the desired Region.

As an analogy, compare maps of Alberta and Brazil. If you are traveling in Canada, you will likely want all the detailed maps of Alberta. However, if you will not be traveling anywhere near South America, then you will probably not need any maps of Brazil. The same is true for our Star-Maps. Each ship can choose which set of maps to load into the database before departure.

The practical process will be as follows. When the star photos are loaded into the Master Database, these photos will be grouped Space Region. There will be a separate file based on each Space Cube Region. The ship can then request photos by collection. The captain will request the Collections for each of the Space Cube Regions of interest. In this way, he will get exactly what he needs, yet only what he needs.

There will also be a computer program which can grab Larger Collections. These are for the Deca-Cube Regions. The captain can simply state he wants the star photos of a specific Deca-Cube Region. The computer will then grab all the files, from all the SCR Regions.

This method can be used for any ship, which travels anywhere in the universe.

Space Regions and Beacon Signal Factors

Another important practical application of the Space Regions is the duplication of the Beacon Signals.

Throughout this publication (and many others) we emphasize that each Space Beacon will emit a unique signal. However, when we consider the larger scales of space, this will not necessarily be the case.

This is where the Deca-Cube Regions become useful. We will establish a rule that within a Deca-Cube Region, each Beacon Signal will indeed be unique. No two signals will be the same.

However, when comparing two or more Deca-Cube Regions, we may use the same Unique Beacon Signal. The reason we can do this is because of distance. When the Beacons with the same "unique" signal are this far apart, there will be no confusion. The ship in one Deca-Cube Region is not likely to be traveling in the other Deca-Cube Region, during the same journey.

This will allow us to (eventually) use some of the same combination of factors. As we place more and more beacons in space, we will eventually be repeating ourselves with the three signal factors. Yet by organizing the Unique Signal per each Deca-Cube Region, we can be assured that all the Space Beacons within a Region will emit a unique signal.

Space Regions and Beacon Maps

This leads us the to the Beacon Maps. First, let us make this clear: All Space Beacons will be in the Beacon Location Database.

Unlike the Star-Map Room, where we selected collections of photos, with the Beacon Location Database each ship will have a full database of all Beacons in the galaxy. This is extremely important for navigation.

Regarding the Beacon Signal, the Beacon Signal will be unique for each Deca-Cube Region. However, there may be more than one Beacon Number, in all of the Deca-Cubes, with the same "unique" signal. In this case, the Computer will display the options for matching Space Beacons, with the Deca-Cube Number alongside the Beacon Number.

The Navigation Crew knows which Deca-Cube Region they are traveling. Therefore they know to simply select that Beacon Number. The Beacon Map will then open up as usual, with that Beacon as the center.

Using the analogy above, this is like getting the same lighthouse identification for both Alberta and Brazil. Yet we are in Alberta, not Brazil. Therefore, we know which one applies to us.

Regional Directors of the Space Corps

The final practical application of the Space Regions is to create the job of Regional Directors of the Space Corps. As we settle space, we will eventually want regional oversight, on behalf of the Space Corps. Therefore, we divide the Universe into Regions, and establish a Regional Director, for each of the Regions.

The Office of Regional Director will primarily be for each of the Single Space Regions. These are the cubic regions of space which are 3 light-years on each side. The Office will physically be located on the most civilized planet in the Region.

Note that we will only need Regional Directors for those Regions where there is civilization. Further note that the Regional Directors will be living on the planets, in those regions. They are not at the Corps Headquarters.

Specific Duties and Organization of Regional Managers

The Regional Manager for the Space Corps will act on behalf of the Space Corps, for that Space Region. The Office of the Regional Manager will be located on the most civilized planet in that Space Region.

The Regional Manager will provide general oversight for all activities in the Space Region. His role is to provide general oversight of all space exploration and space settlements in the Region. He will not dictate, but rather be the oversight and final decision-maker when needed.

The "Manager" will begin as a single person, then gradually develop into a leadership staff with an executive manager. The Regional Manager will also communicate regularly with other Regional Managers in neighboring Regions, as well as with the Corps Headquarters as determined necessary.

Using the Universe Grid System
with Space Object Location Coordinate System

It is now time to compare the two Location Systems discussed in this publication. The systems are meant to be used separately, yet they can be sometimes used together. Generally, the distinction is the following:

1) We will look at the Space Regions (of the Universe Grid System) to consider each Region as a whole. 2) We will look at the Direct Path (of the Location Coordinate System) to travel to a specific destination.

Comparison of the Two Systems

The Universe Grid System is really a separate coordinate system from the Space Location Coordinate System. The Universe Grid System is designed to consider Regions of space and everything in that Region.

However, the Space Location Coordinate System is designed to locate specific objects in space. This System will tell you the Direct Path, with Direct Angle and Direct Distance, to any object in space.

1. When you want to study all the stars, planets, and beacons in a Region of Space, then use the Universe Grid System.

2. When you want to locate a specific object in space, then use the Space Location Coordinate System.

Space Object Location Coordinate System: Review

Let is review the Space Location Coordinate System. This Coordinate System is the coordinate system for precisely locating any object in space.

All objects in space are located using a Direct Angle and Direct Distance. The Direct Angle has two values: X-Y Angle, and the Z Angle. Earth is of course the official reference point. Therefore, using this system, all objects in space can be precisely located. These are the Official Locations for each star, planet, and beacon in the galaxy.

Universe Grid System: Review

The Universe Grid System is designed to organize the Regions of Space. These Regions are useful for considering an entire Region at a time. This includes obtaining proper maps for that Region. It also includes ensuring the Beacon Signals are unique within each Region; yet can be the same from Region to Region.

Simple Example of Showing Both Systems

Generally, we will use either one system or the other. This is because the purposes are different. However, we can show how these two systems overlap in physical space. The illustration below is a simple example.

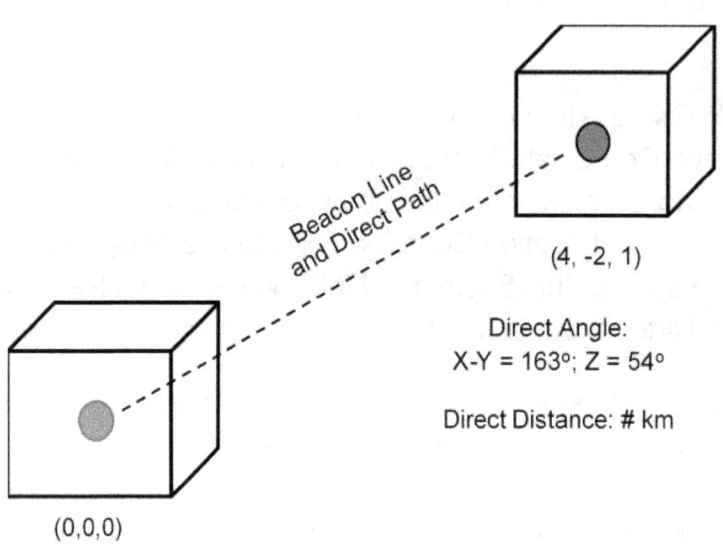

Beacon Line and Direct Path

(4, -2, 1)

Direct Angle:
X-Y = 163°; Z = 54°

Direct Distance: # km

(0,0,0)

Chapter 17:

Review of Navigation Systems

Navigation Systems are Essential for Space Travel

Humans will be exploring space. There is no doubt that this will be our future. We will indeed explore and settle various planets through the galaxy. We will eventually reach beyond the galaxy as well.

However, in order to make this endeavor realistic, we require certain technologies and infrastructure. This includes navigational systems. It is for this reason that the authors of this publication have set forth the best systems possible for navigation across space.

The first system we must use is the Coordinate System. The existing coordinate systems are inadequate for realistic space exploration. Therefore we have developed a better system for determining the locations of objects in space. It is important that we establish this system, and use it from the beginning.

Coordinate System for Space Locations

Before we can develop any navigation system, we must first establish a coordinate system. We must have a method for locating any object, in any location in space. Therefore the establishment of the Coordinate System for Space Locations is extremely important.

The existing location systems are inadequate for space exploration. Therefore, the authors studied the science of mapping and astronomy (as separate disciplines and as combined sciences) to develop the best coordinate system for our needs. Using this Coordinate System, we will be able to accurately locate the position of any celestial object. The system is briefly reviewed below.

Reference Point: Earth

We begin with specifying a reference point. This reference point will be Earth. Thus, the position of Earth will serve as the zero distance, as well as the 0 degrees in all directions.

However, each Home Planet will want to know locations based on its position. Therefore, the computer programs will be able to automatically convert the Official Location to the Location as based from a different Home Planet. The user simply selects a different Home Planet name, then tells the computer to perform the calculations.

Yet the Official Location of objects will always be from Earth. The other Home Planets are simply able to perform the conversion calculations.

Direct Angle

The Direct Angle is the most direct path between the Home Planet and the Object of Interest. The Direct Angle is best given as two values. These are the X-Y Angle and the Z Angle.

X-Y Angle; Simply X Angle

The X-Y Angle is essentially a circular graphing system, similar to a compass. The X-Y Angle is based on a horizontal plane. We will impose this horizontal plane on the Earth's Orbit. We will also specify that the zero degree will be where the Earth is closest to the Sun in the Orbit. The sequential degrees will then proceed in the direction of the Earth's Orbit.

Notice that because this a circular graph, we can also refer to this simply as the X-Angle.

Z Angle

The Z Angle is based on the equator of the Earth. The zero degree of Z Angle is the Equator. Therefore, a positive Z angle is an angle upward from the equator, as if toward the North Pole. The negative Z angle is then downward, as if toward the South Pole.

Zero Distance: Earth's Position as Closest to Space Object

The Zero Distance is the location of the Earth. However, the Earth does orbit the sun. Furthermore, the destination planet can be in any direction from our solar system. Therefore, we must specify a location of the Earth, in its orbit, for the exact zero location.

Thus, the Exact Zero Distance is the location of the Earth, where the Earth is closest to the destination solar system. This is most practical, as it is from this location in the Earth's orbit where all ships will be leaving, when heading to that solar system. Therefore, for practical interest of the ships, this location is best for the zero distance.

Of course this means that different Earth Orbit Locations will be chosen, as the starting point for different destinations. Yet for practical purposes of space travel, this is the most sensible option.

Universal Zero Distance

The Universal Zero Distance is the location of the Earth where the Earth is closest to the Sun. This location will therefore be consistent for all distance measurements, to any location in space.

However, the Official Zero Distance will always be the location of the Earth described above. We will simply have two tables of distances: the first based on the Earth's Closest Position to Object; the second based on the Earth's Position when Closest to the Sun.

Direct Distance

The Direct Distance is the distance between Home Planet and Object of Interest. The Direct Distance is one of the two Official Location Factors to known the location of any object.

The Object of Interest is usually a Destination Planet, Space Beacon, or Star. The exact distance is measured using the Path Drawing Program.

Precision in Direct Distance

The precision in the Direct Distance depends on the Object of Interest. For the Space Beacon, the beacon rarely moves. Therefore the Direct Distance is known absolutely.

However, the Destination Planet moves in its orbit. Therefore, we can use methods to improve the exact distance. For the Destination Planet, we use the same criteria above. That is, where the Destination Planet is closest in its orbit to the Home Planet, this becomes the Official Location for the Direct Distance.

The stars also move through space, though at slower pace. The distance values will generally be accurate enough, regardless of the star's movement. However we can (and should) adjust the distances to the stars every 20 years or so.

Official Location: Direct Angle and Direct Distance

The Official Location, for all objects in space, will be given by the Direct Angle and the Direct Distance. The Direct Angle is the most direct path from Earth to the Object of Interest. The Direct Distance is the shortest distance, between Earth to the Object.

Therefore, using the specific definitions for the Coordinate System as explained above, we can give a Direct Angle and Direct Distance for all objects in the galaxy. We can therefore state with precision the location of any planet, any beacon, or any star throughout the universe.

Conversion Systems from Different Home Planets

Note that each Home Planet will have conversion programs. These programs will convert the Angles and Distance, to each object, based on their Home Planet.

However, the Official Location will always be from Earth. This is necessary, to maintain consistency in all databases.

Using this Official Location System from Beginning of Exploration

This Coordinate System is the best system available for specifying the location of objects in space. We must therefore use this system for our exploration and settlement of space. We will use this for the locations of all destination planets and all space beacons.

It is also very important that we use this this Location System from the very beginning of our Space Exploration. It is important that we establish consistency now, so that our databases and mapping systems will be universally effective in the future.

Specific Navigation Systems

Having established the Coordinate System, and the method of specifying the precise location of any object in space, we can now review the Navigation Systems.

In general, a Navigation System must be able to answer the following questions: Where are we now? Where is our destination? How do we get from here to there?

For the vast distances of space, these questions can be difficult to answer using traditional navigation methods. Therefore it is important to develop the best Navigation Systems possible for Space Travel. We will now review the best of the Navigation Systems which we have developed for this purpose.

These Navigational Systems can be grouped into the following:

- Traditional Navigation Methods

- Path Drawing Program

- Star-Map Room

- Space Beacons and the Beacon Map

Traditional Navigation Methods

The Traditional Navigation Methods involve visual sightings and using tools such as the sextant. These are the methods used by the ocean ships for millennia, and to a certain extent, also used by astronauts in recent times. However, these methods are limited for long distance space travel.

The basic method for traditional navigation is orient your ship relative to a certain group of stars. Keep one group of stars to your left, another group of stars to your right, and a third group of stars straight ahead. Use the sextant to note more precise angles of your course.

Although approximate, this method can be used as a general navigation guide. The crew will be able to observe the stars as they travel, and note any deviations which seem unusual.

However, this method is only approximate. Furthermore, any slight deviations in space travel can lead the ship far from course. Therefore the other navigational systems are designed to be more precise.

Path Drawing Program

The Path Drawing Program is a navigation system which will track the exact paths taken by the ship. From this information, the Program can draw a Direct Path from Home Planet to Current Location.

The Path Drawing Program is primarily used by the Beacon Placement Ships, to know the exact location of each beacon they put into position. However, variations of the Program can be used by other ships.

Mechanical Engineering Becomes Data

The process begins with the mechanical engineering of the ship. Every time the ship changes course or speed, this information is recorded in the computer. The computer will therefore track each path taken, while noting the speed and duration of time. These will become Path Steps, which can be drawn on the computer.

Events and Path Steps

An "Event" is defined as any time the ship changes direction or speed. Regarding direction, this can be changes in X, Y, or Z direction. Changes in speed, for the same course, will also become an "Event".

A "Path Step" is then the specific path taken by the ship, between any two Events. Thus, each Path Step is the direction, speed, and total time the ship takes; before changing direction or speed again.

On the Mapping Program, the Events are shown as dots. The Path Steps are shown as segments between the dots. In this way, the navigation crew will have a visual representation of each Path Step of their journey. Note that this information is automatically updated every 8 hours.

Direct Path and Exact Location

The Direct Path is the most direct path from the Home Planet to the Current Location. The ship may have changed course from time to time. The ship will likely have changed speed from time to time. Yet from these changes, we can calculate the most direct route to their location.

The computer will take all of the individual Path Steps, then perform the appropriate calculations, to determine the Exact Location of the ship, at that moment. This location will be very precise.

The computer will therefore be able to determine the Exact Location, and the Direct Path between Home Planet and Current Location. This will be used for the Official Beacon Locations.

Official Beacon Locations

The primary purpose of the Path Drawing Program is to determine the Exact Location of the Space Beacons. The Placement Ships will have the most sophisticated versions of the Path Drawing Program, and will therefore know the Locations very precisely.

Thus, when the Placement Ship is ready to place the next Beacon, the crew will perform the final calculations of the Path Drawing Program. The Program will determine the Direct Path and Exact Location. The Program will therefore provide the Official Direct Distance and Official Direct Angle (X and Z Angles) for the Beacon Location.

Sophistication of Program, Accuracy, and Other Ships

The accuracy of the Path Drawing Program will depend on the sophistication of the components. This includes the precision of the mechanical engineering and measuring devices. This also includes the mathematical computations in the computer programs.

However, the more advanced components are more expensive to create, and require precision in assembly. Therefore, other ships may choose to not have this level of sophistication. Their Path Drawing Program will certainly be more accurate than just visual sighting, yet not absolutely precise.

The Beacon Placement Ships, however, must always have the most sophisticated and most precise components for the Path Drawing Program. The Beacon Placement Crews are essentially the Surveying Teams of Space. Therefore, they must have the most accurate equipment.

Star-Map Room

The Star-Map Room is an immersive display of stars. While in this room, the viewer will experience the environment as if he were truly at the location in space.

The structure of the Star-Map is a dome, with individual screens. Each screen shows a specific region of stars. Taken together, all of the screens will show an entire view of space, at any given location. Thus, you are completely immersed in the reality of space, while in this room.

The images themselves are photos taken from ships which previously traveled in the region. Thus, the viewer is indeed seeing the reality of space, at that location.

Star-Map Room Structure

The Star-Map is a dome structure, built within a square room. Thus, the Dome Display is a stand-alone structure within the square room. It is designed to hold the display screens, not support of the outer room.

Note also that the user will enter through two doorways. The first is to the square room. The second doorway is to the Dome itself.

The structural grid of the dome will not only create the Dome, this will also support the image screens. The holes between the grid of the structure will be the location of the screens. The screens are nested gently in trays, which face into the room.

Furthermore, the structural supports will also house the electrical wires and data cables. The electrical wires will power the screens. The data cables will bring the specific star photo to the appropriate screen.

Screen Displays

The immersive reality is based on the collection of screen displays. These are flat screen displays, which sit in trays, and face toward the room. Each screen will show a specific photo. The collection of these photos, together, will create the immersive reality of being in the location.

This is similar to a puzzle, where all the pieces are put together. Though each piece is distinct, the entire collection of pieces creates a much larger scene.

The specific photos are taken from ships which have previously traveled at the location. Therefore, what the user sees in the Star-Map Room is indeed the actual scene at this location.

Star-Map Room: General Operation

When a person desires to use the Star-Map Room, he walks to the pedestal in the center of the room. This is the location of the computer.

The user activates the computer. He selects two reference objects. These reference objects are stars, planets, or beacons. The user then specifies which sides of the Dome Display to show those reference objects. The computer then automatically determines how the Dome Display should look, based on those criteria.

The computer will then call up all of the specific photos from the database, to make the Display appear as it should. The computer then sends each photo to the appropriate screen. This produces the immersive reality, as if the person was actually standing in that location in space.

Database of Photographs

The database for the Star-Map Room will be a comprehensive database, of all photos taken in space. From this database, the users can recreate the scene anywhere in space. The crew of this ship can therefore see the reality of any location, which has been traveled by a previous ship.

The database, for each ship, will of course improve over time. Each ship will take additional photos. These photos will be transferred to the Master Database on each planet. The database on each next ship, before embarking, will be updated with the latest photographs.

Thus, the databases for the Star-Map Room, on each ship and each planet, will become more complete and more accurate over time.

Beacons, Supply Containers, and Planets

The most common needs for the navigation crew will to be looking for any one of the following: Space Beacons, Supply Containers, and Settlement Planets. Each of these can be entered as Reference Objects. The Star-Map will be created around these objects.

Indeed, most of the photos will be taken at those locations. Therefore the accuracy of the photos will be greatest for these objects. Thus, the crews will know exactly how the stars should look, outside their window, when reaching these locations.

The crew can also ask the computer to locate these objects. The user can ask to recreate the display from their current location, then ask the computer to find the nearest Beacon or Supply Container. The computer will then highlight those points on the screens. Data regarding each can also be displayed on one of the adjacent screens.

<u>Star-Map Room on Any Ship or Planet</u>

This Star-Map Room can be used on any ship, and on any planet. There are only two variations: the size of the room, and the number of photos in the database.

For larger locations, such as on a planet or large ship, then the Dome Display can be quite large. However, for smaller ships, we may need to make this Dome Display much smaller.

The number of photos may also vary. With more photos available, we can have greater accuracy in the Dome Display. However, this requires greater computing power. Therefore smaller ships may want fewer photos.

We can also create Collections of Photos. Each "Collection" is a group of photos for a Region of Space. Any one ship may only want the Collections of Photos which apply to their region of travel.

<u>Computer Programs</u>

The computer programs for the Star-Map Room must also be able to perform sophisticated functions. There are several important functions.

1. The computer must first be able to recreate any scene, within the computer itself. This is the scene which will then be shown in the Dome Display. Thus, the computer must be able to take any two points, as selected by the user, to create the scene as it should look on the Dome.

2. The computer must then be able to select the specific photos which will create the Dome Display. The computer will be able to search through the entire catalog of photos, to select the individual photos, which will create the Scene.

3. The computer will then be able to send the specific photo to each display screen. That is, each Screen will be sent the proper photo. This will create the entire Dome Display, and Immersive Reality.

Star-Map is Always Improved Over Time

The Star-Map is always being improved over time. As each ship travels through space, the crew will take photos (according to Corps Protocol). These photos are then brought to the Master Database, and then uploaded to each next ship before embarking.

Therefore, the Star-Map Database, for each embarking ship, will have the most comprehensive collection of photos, The crews will be able to see many more regions of space, and view these locations from different perspectives. This process is ongoing, and always improving.

Space Beacon as Navigation Guide

The Space Beacons are navigational guides and communication relays, placed in a series, between every two Settlement Planets.

As Navigational Guide, the Space Beacon acts as lighthouse and mileage marker. Each Space Beacon emits a unique signal, which then be correlated to a location on the Beacon Map. Therefore, using the Space Beacons the ships will know their exact location in space.

Space Beacon as Lighthouse

Each Space Beacon will emit a burst of electromagnetic energy. This similar to a lighthouse. The passing ship will receive this signal. Furthermore, the cycle time of the EM burst will be unique to that Beacon. Therefore the Beacon has a unique signal, and can be located on a Beacon Map.

Unique Beacon Signal

The Unique Beacon Signal is based on three factors. These are 1) the frequency of EM emitted; 2) the amount of time burst is emitted; and 3) the amount of time between bursts. These three factors, together, will make each Beacon Signal unique. Notice that some beacons may have the same values for one of the factors, but not for all three.

Thus, we can be assured that no two beacons will emit the same signal. Each signal is unique.

Knowing Precise Location of Space Beacon

The Space Beacons will have a precise location. This location will also remain constant throughout time.

We know the precise location because the Measurement Crews of the Beacon Placement Ship will perform the specific calculations. The primary tool is the Path Drawing Program, which will determine the location with great accuracy. The Measurement Crew will take other measurements as well. These become the Official Location Data for the Beacon.

The Space Beacon will also remain in place because there is object nearby to cause gravitational influence. In addition, many of the Beacons will have the Automatic Self-Correction Systems. These systems will periodically determine shifts in location, and make necessary adjustments.

Therefore, we will know the location of each Beacon with absolute certainty, and this location will remain constant essentially forever.

Beacon Database

The Beacon Database will have the complete data of Beacon Number, Beacon Location, and Unique Signal, for all Space Beacons in the galaxy.

The Master Database is kept on Earth. Duplicate databases will be on each planet, and on each ship. These databases will of course be updated, as each new Beacon has been placed into position.

Beacon Map

The Beacon Map will show the location of any Beacon, throughout the galaxy and beyond. The Beacon Map will therefore be one of the ultimate navigation tools for space travelers.

When a ship passes a Space Beacon, the receiver will analyze the Beacon Signal, and match to the specific Beacon Number. The Navigation Crew can then open the Beacon Map, to see the Beacon displayed. The user can of course perform various zoom options. This will allow the user to compare the Beacon Location to any star or planet. Thus, the Navigation Crew can see the position of that Space Beacon, relative to any star or planet. This also means that the Navigation Crew will know their own exact location, relative to any planet or star.

Data regarding the Space Beacon is also displayed on the side. The menu on the side will also allow the user to search for other data regarding that space beacon, or to search for other beacons.

Therefore, using the Beacon Signal and the Beacon Map, the crew will know their exact location. They will also know the distance to the next Beacon, and remaining distance to their destination planet.

Beacon as Communication Relay

The Space Beacons also serve as Communication Relays. This is important for sending messages across the vast distances of space.

As a Communication System, the Space Beacon will receive the message, boost the intensity, then send the message onward. Furthermore, there are many of these Space Beacons in a series. Therefore, the signal is boosted at periodic distances. In total, the series of Space Beacons, as Communication Relays, will ensure that all messages are received at the destination planet with significant strength. In this way, messages can be sent between any two planets, with great clarity regardless of the distance apart.

Using Beacon for Ship-to-Planet Communications

The ships which travel through space will also use these same Beacons as Communication Systems. The dishes on the side will receive the messages from the ships, then send through the Beacon Line, to the destination planet. Planets can send reply messages to the ships using the same system.

The communication system is also an important part to ensure safe travel. All ships will be required to send status messages, at each Space Beacon. This will allow the staff at the home planet to know the location of the ship, and the general status.

This will allow the staff at the home planet to be assured that the crew and passengers of the ship are on course, and safe. If a ship does not report when expected, a rescue ship will proceed to the latest known location. Furthermore, if a ship requires assistance of any kind, the home planet will notify other ships nearby to offer assistance.

Powering the Space Beacons

The Space Beacons will be powered using nuclear power and solar power. Nuclear power is effective, because a small volume of nuclear material will provide enough power to last many years. The supply will need to be replaced, but this can be done on a known schedule.

Solar power is effective where the beacon is close to a bright star. The solar power will recharge the batteries, and provide continuous power for the Beacon. This is best for beacons in a solar system.

Particle Supply for Space Beacon Operations

In order to emit the Beacon Signal, we must first have neutrinos or photons from which to make our Beacon Signal Photons. We will collect these from the background space, as well as the brightest stars nearby.

It is essential that we design the Particle Harnessing Systems into each of the Space Beacons. Without capturing and converting these particles, regularly and in large enough quantities, the Space Beacons will be completely useless.

There are several methods for capturing these supply particles. These methods are discussed in the publication on Particle Harnessing.

Specialized Beacons and Other Equipment

The Beacon Line contains not only the Main Space Beacons, but other types of beacons as well. These include the Solar System Beacon and the Node Beacon. The Solar System Beacons are essentially the same as the Main Space Beacons, yet located within a solar system. This means the Beacon is affected by the gravity of the central and other planets. Yet this also allows us to use that central star for solar power and particle supply. Thus, we must make several technical modifications for the Solar System Beacons that are not needed for the Main Space Beacons.

The Node Beacons are located at the Nodes, which are the intersection points of several Beacon Lines. The Node Beacon will manage all communications from each Beacon Line, as well as being able to transfer messages from one Beacon Line to another.

The Beacon Line also includes the Surface Antennas and the Satellites above each planet. The Surface Antenna transmits messages from planet to the Satellite above, which then sends the message to the nearest Space Beacon. Messages are received in a similar way: the nearest Beacon sends the message to one of the Satellites, which then transmits the message down to the Surface Antenna.

Maintenance of Space Beacons

The Space Beacon should not require much maintenance. The Space Beacons are designed to operate independently, and automatically, for long periods of time. This is necessary for their positions in deep space.

However, they will require some maintenance. The most common maintenance will be to replace the Power Supplies. This will be done on a regular schedule, and all Beacons along the same Beacon Line will have their power supplies replaced on the same maintenance trip.

The maintenance of Space Beacons will be performed by specially trained Beacon Maintenance Crews.

Note that replacing the Power Supplies and other Cartridges can be done in space, without moving the Beacon. The Space Beacons are designed to make these operations easy.

However, some maintenance will require the beacon to be brought to the ship. This will only be done on specially designed Beacon Maintenance Ships. Furthermore, there is a very strict protocol for moving the Beacon. This is to ensure that the Beacon is placed in the precise location once again.

Chapter 18:

Glossary by Topic

Overview of Glossary

The following sections are short definitions of some of the most important terms regarding Space Navigation. The terms are organized by general topic, then listed in logical order within the topic section.

The Topic Sections for the Glossary are as follows:

    A. Coordinate Systems

    B. Universe Grid System and Space Regions

    C. Navigation Systems

    D. Path Drawing Program

    E. Star-Map Room

    F. Space Beacons

    G. Beacon Signal

    H. Particle Supply

Note that the definitions are short. The full explanations for each term are found in the text of the publication.

A. Coordinate Systems

Space Location Coordinate System

The Space Location Coordinate System is the coordinate system for precisely locating any object in space. Some of the main concepts include:

- Earth as Reference Point
- Horizontal Plane (X-Y) as Earth's Orbit
- Vertical Plane (Z) the Axis of the Earth's North-South Pole
- Official Location as Direct Angle and Direct Distance
- Conversions Available for Each Settlement Planet

Official Location

The Official Location, for any object in space, is designated by the Direct Angle and the Direct Distance, as discussed throughout the text.

Direct Angle

The Direct Angle is the angle from Earth to the Object of Interest. This is given as two angles: X-Y horizontal circle angle, and Z vertical angle.

The horizontal plane is imposed on the Earth's orbiting plane. The zero degree is where the Earth is closest to the sun. The vertical plane is based on the Earth's North Pole-South Pole Alignment, with the zero degree at the Equator

Direct Distance

The Direct Distance is the straight line distance from Earth to the Object of Interest. Note that there are two Direct Distances: Official and Universal. The Official Direct Distance begins where the Earth is closest in its orbit to the Object of Interest. The Universal Direct Distance begins where the Earth is closest to the Sun. The Official Direct Distance is more practical for Beacon Placement and all ships traveling the route.

B. Universe Grid System and Space Regions

Universe Grid System

The Universe Grid System is the method for dividing the entire universe into uniform regions of space. Each region can be numbered and placed on the Universe Map.

The Universe Grid System, with the Space Regions, allows us to organize the physical areas of the universe. This system will allow us to have better scientific discussions, improve the practicality of space travel, and provide general oversight.

Space Regions

The Space Regions are the uniform regions of space, within the Universe Grid System. The basic Space Region is a cubic region, with dimensions of 3 Light-Years per side. Larger Space Regions can be designated as multiples of the Single Space Region.

Each Space Region is located by a Prefix and a set of Step Numbers. The Prefix tells us the size of the Region; the Step Numbers tell us the position of the cube in the Universe Grid.

Step Numbers

The Step Numbers are the three values which tell us the position of the Space Region within the Universe Grid System. The Space Regions exist as cubic regions, similar to blocks. These blocks are stacked together, throughout all of space. Thus the Step Numbers will designate the exact location of the Space Region (the cube) within the stack.

The Step Numbers are given as three values. These are for the X, Y, and Z directions. Starting from the First Cubic Region (which includes Earth), we count blocks (the Step Numbers) for each of the three directions. We have thus located the specific cubic region within the Grid.

Single Space Cubic Region (SCR)

The Single Space Cubic Region (SCR) is the Base Region Size for the Universe Grid System. The Single Space Cubic Region has dimensions of 3 Light-Years per side. Each region is designated by the Prefix "SCR", followed by the three Step Numbers.

Note that this Region can be referred to simply as the Space Region, the Space Cubic Region, or as the Single Cubic Region.

Deca-Cube Region (DCR)

The Deca-Cube Region (DCR) is the next larger region size, in the Universe Grid System. The Deca-Cube Region is a larger region, made of numerous Single-Cube Regions.

Specifically, the Deca-Cube Region is 10 Single Cube Regions, on each side (X, Y, Z). The total dimension of the Deca-Cube Region is therefore 30 Light-Years on each side.

The Deca-Cube Regions are located similar to the Single Cube Regions, using the Step Numbers. However, a different prefix is used, to distinguish the different Region Size. Thus, the Deca-Cube Regions are located with prefix "DCR" followed by the three Step Numbers.

First Space Region

The First Space Region is the Center Cube in the entire Universe Grid System. For practical reasons, our planet Earth is in the center of the First Space Region. All other Space Regions are placed around this cube.

The First Space Region is given the Step Number of "0,0,0". All other cubic regions are located in Steps, based on this First Region.

Regional Manager for the Space Corps

The Regional Manager for the Space Corps will act on behalf of the Space Corps, for that Space Region. The Office of the Regional Manager will be located on the most civilized planet in that Space Region.

The Regional Manager will provide general oversight for all space exploration and space settlements in the Region. He will not dictate, but rather provide oversight and be final decision-maker when needed.

The Regional Manager will also communicate regularly with other Regional Managers in neighboring Regions, as well as with the Corps Headquarters as determined necessary.

C. <u>Navigation Systems</u>

<u>Traditional Navigation Systems</u>

The Traditional Navigational Systems are those used by ships for centuries. This includes line of sight, alignment with specific stars, and use of the sextant.

<u>Path Drawing Program</u>

The Path Drawing Program is a set of tools, which will track all motions of the ship, and therefore be able to plot every path taken by the ship, up to that moment.

The Mapping System of the Path Drawing Program can show the individual steps taken during the travel. The Map can also compute the Direct Angle and Direct Distance, from Home Planet to Current Location.

<u>Star-Map Room</u>

The Star-Map Room is a Dome Structure which displays photos of the stars. The result is an immersive experience, as if being in that location. The specific view can be selected, based on specific stars or planets.

<u>Space Beacons</u>

The Space Beacons are modest sized equipment, set in a direct path between two planets. The Beacons act as Lighthouse, Mileage Marker, and Communication System. The location of the Space Beacon is precisely known, and will remain in that location for all time. Therefore the Space Beacon can be matched to a Beacon Map, and the ship's crew can determine their precise location.

D. Path Drawing Program

Path Drawing Program

The Path Drawing Program will track the path taken by the ship, and can draw this path on a computer screen. The "Program" is actually a set of tools, including mechanical engineering devices, measurement devices, databases, calculation programs, and mapping programs.

The Path Drawing Program is primarily used to determine the Official Location of Beacons during placement.

Event

An "Event" in the Path Drawing Program is any time the ship changes direction or speed. The ship must be in steady state of that direction and speed for 20 minutes, for the computer to note this as a new Event. The Event is labeled with X and Z angles, speed, and time stamp.

Path Step

The "Path Step" is the path of the ship in a specific direction. The Path Step is computed between the two "Events". On the Mapping Program, the Path Steps appear as individual segments.

Direct Path

The "Direct Path" is the most direct route from Home Planet to the Current Location. The computer will take each of the individual Path Steps, perform the necessary calculations, and then produce the result.

The Direct Path is shown on the Map as a straight line from the Home Planet to the ship's current location. The values are given for the Direct Angle and Direct Distance. These values can become the Official Location Data for the next Space Beacon.

E. Star-Map Room

Star-Map Room

The Star-Map Room is an immersive display of stars. The display itself is a dome structure, made of individual screens. Each screen shows a specific photograph taken from previous ships. Taken together, the person in the room will experience the reality as if he were there.

Dome Display

The Dome Display is the entire scene which surrounds the viewer. The Dome Display is composed of individual Screens. This is similar to a puzzle, where the individual pieces create a cohesive scene.

The term Dome Display refers to the total scene that is created from the individual photos. The term also refers to the completed structure, of dome grid and display screens, without any photos being displayed.

Display Screen and Screen Trays

The Display Screen is a screen which shows the specific star photos. The Display Screen is essentially a flat screen, which rests in a Screen Tray. These Screen Trays are connected to the grid lines of the dome structure. Each Display Screen rests inside the Screen Tray.

The power and data cables are housed inside the grid structure. Each Display Screen is then given its specific photo, to create the entire Dome Display.

Star-Map Database and Computer System

The Star-Map Database is the collection of star photos which will create the Dome Displays. The individual photos have been taken by all previous ships during their journeys. The database of photos become more comprehensive over time, as photos of different regions and different angles are added.

The Computer System for the Star-Map Room will perform all the functions needed to create the desired immersive display. The user will select the reference points (stars or planets) then the computer will determine the entire set of photos required for the display. The computer will then call up the needed photos from the database, and send specific photos to the appropriate Screen Displays.

F. Space Beacons

Space Beacons

The Space Beacons are permanent fixtures in space which serve as navigation guides and communication relays. The placements of the Space Beacons are precisely known, and will never change.

Beacon Line

The Beacon Line is a series of Space Beacons between two planets. Each pair of settlement planets will have its own Beacon Line.

In practical use, the term "Beacon Line" has several applications:

- The physical series of Beacons between planets
- The communication line between planets
- The route taken by ships traveling between planets.

Beacon Network

The Beacon Network is the entire network of space beacons throughout the galaxy. The Beacon Network consists of each Settlement Planet and the Beacon Lines which connect the planets.

The Beacon Network will be the communications system throughout the galaxy. The Beacon Network will also be the primary navigation system for space travel. The Network will grow and evolve over time.

Main Space Beacon or Deep Space Beacon

The Main Space Beacon, also known as the Deep Space Beacon, is the primary Beacon Style in the Beacon Network.

These Beacons are placed beyond the boundaries of each solar system. Thus, these Beacons are also Deep Space Beacons, as they are the primary beacons of the Beacon Line in deep space.

All other Beacons are based on the design of the Main Beacon.

Solar System Beacon

The Solar System Beacons are those Space Beacons which are placed within a solar system or at the edge of a solar system. Their positions are gravitationally affected by the central star. Yet this same star can also provide solar power and supply particles. The primary benefit of the Solar System Beacon is to give messages a final push of intensity.

Node Beacon

The Node Beacon exists at the intersection points (nodes) of multiple Beacon Lines. The Node Beacon is able to manage multiple messages arriving from different locations, at the same time. The Node Beacon can also transfer messages from one Beacon Line to another.

The Node Beacon is generally the furthest from any solar system. Therefore the power supplies must be able to provide power for much longer periods of time.

The Node Beacon will also require a greater amount of supply particles. This is necessary to relay the messages and to emit the beacon signal in multiple directions.

Surface Antennas and Geostationary Satellites

The Surface Antennas are the antennas on the ground of the settlement planet. The actual antenna can be of any design, including individual antenna, dish, or array of dishes. The purpose of the Surface Antenna is to send messages from planet surface to satellite, and to receive the messages from satellite to planet surface.

The Geostationary Satellite sends the messages from the surface antenna to the nearest beacon. The Satellite will similarly receive messages from the beacon, and transmit to the surface antenna.

G. Beacon Signal

Beacon Identification Signal

The Beacon Signal is the unique signal emitted by each Space Beacon. The Beacon Signal is a burst of EM energy, of specific frequency and cycle time. This signal is unique to each Beacon, and can therefore be correlated to a specific Beacon Number. This allows the Beacon to be located on the Beacon Map, and confirm the position of the passing ship.

Unique Signal Factors

The Unique Signal Factors are the combination of factor which make the Beacon Signal unique for each Beacon. The three factors are:

- Frequency of EM Burst
- Amount of Time the Burst is On
- Amount of Time the Burst is Off

These factors are listed in the Beacon Design Database and the Beacon Location Database. The staff at Corps Headquarters will ensure that no two Beacons, in the same Region, have the same set of 3 factors.

Beacon Location Database

The Beacon Location Database contains all the essential location data for each Space Beacon. This data includes: Beacon Number, Beacon Signal Factors, Official Beacon Location, and other factors. This information will lead to the Beacon Map.

The Beacon Database will be updated as new Beacons are placed. The latest version will be uploaded to each ship before embarking.

Beacon Location Map

The Beacon Map will show the exact location of any Space Beacon, relative to selected stars or planets. The Official Location of the Beacon is measured in reference to the Earth. This data is in the Beacon Location Database. The user of the database can call up any Space Beacon, either by Beacon Number or Beacon Signal. The Map View can be adjusted by selecting a reference star or planet.

Using this Beacon Map, the user will know the precise location of every Space Beacon. The user will also know their own location, when they pass one of these Beacons.

H. Particle Supply

Particle Supply

The Particle Supply is the collection of particles needed for Beacon Operations. The specific particles will be neutrinos or photons. These Supply Particles are needed for Beacon Signals and Boosting Messages.

*Note that without a regular supply of these particles, the Space Beacons will not be able to perform their intended functions.

Neutrino

The neutrino is a small particle which is packed with enormous amount of energy. The neutrino is sealed, which ensures that the energy is contained. Thus, the neutrino will travel at the same speed forever.

The neutrino is also the base particle for the Photon System. All of the particle aspects of the photon system are due to the neutrino. Therefore, we can capture the neutrino and convert to the photon system of desired frequency.

The neutrino does have mass, and gravity. Mass exists in the form of mass spots on the surface. Gravity exists as wisps of gravity strings which extend from the mass spots. These gravity strings hold the neutrinos together in clusters when first emitted from a star.

Photon or Photon System

The Photon is more accurately called a Photon System. The Photon System is essentially a high-speed train, where the neutrino train carries the EM energy string passengers.

The Photon System begins with the Photon Core, which is also the Neutrino. Then the EM energy strings are attached to the Neutrino.

The number of energy strings determines the total EM energy of the photon system. The total EM energy then determines the frequency of pulsation. The wavelength is a combination of the forward speed of the neutrino with the pulsation frequency of the EM strings.

Photon systems can exist as many possible energies (and therefore many possible frequencies). We can capture many of these photon systems from space, then convert them to the desired photon frequency of the Beacon Signal.

Supply Particle Harnessing

The process of Supply Particle Harnessing will capture neutrinos or photon systems from space, then convert these particles to the desired photon frequency. The specific method of Supply Particle Harnessing depends on the type of particle, the source of the particle, and where the particle is being captured.

For photon capture, the primary method is to use a set of Capturing Dishes. These dishes capture the microwave energy photon systems from the background of deep space.

For neutrino capture, the primary method is use a Capturing Sheet. The Sheet is essentially a grid of wires, where the electrons absorb the neutrinos from space, then channel to storage.

Supply Particle Storage

Supply Particle Storage is the method for storing the Supply Particles until needed. Notice that due to the regular need of Supply Particles, the Storage will be filled, emptied, and re-filled on a regular basis.

For photons, the primary method for Supply Particle Storage is a box with layers of reflective coating. The reflective material is selected to perfectly reflect the frequencies of the photons captured.

For neutrinos, the primary storage method is to use loop wiring and capacitors. This will keep the electrons contained and moving, which will therefore keep the neutrinos inside the electrons; and inside the box.

Particle Conversion

The process of Particle Conversion is to take the Supply Particle, then convert the particle to a photon system of the desired frequency. This process involves adding or removing a certain number of EM strings.

The neutrino begins with no EM energy strings. Therefore we will add the amount needed to create the desired frequency.

The photon system, as supply particle, already begins with some EM energy strings attached. Therefore we will either add some EM energy strings, or remove some EM strings. This will convert the photon system of as supply particle, into photon system of desired emission frequency.

Additional Publications
in the Space Corps Series

For further details on many on the topics discussed in this publication, please see the other Publications in this series. Each publication expands on one of the topics, with more technical details and illustrations.

These publications include the following. The number refers to Volume Number in the Series.

1. Introduction to the Civilian Space Exploration Corps

2. Organizational Structure of the Space Corps: Abridged

6. Designing Space Ports

10. Space Beacons for Navigation and Communication

11. Navigation Systems for Space Travel

12. Communication Systems for Settlements Across the Galaxy

13. Placement, Maintenance, and Adjustment of Space Beacons

14. Particle Harnessing Systems for Space Beacons

15. Databases and Mapping Systems in the Space Corps

www.ingramcontent.com/pod-product-compliance
Lightning Source LLC
Chambersburg PA
CBHW080824220526
45467CB00008B/2190